VOL. 2

Uncovering
Student Ideas
in Science

25 More Formative Assessment Probes

By Page Keeley,
Francis Eberle,
and Joyce Tugel

NSTApress
National Science Teachers Association
Arlington, Virginia

National Science Teachers Association

Claire Reinburg, Director
Judy Cusick, Senior Editor
Andrew Cocke, Associate Editor
Betty Smith, Associate Editor
Robin Allan, Book Acquisitions Coordinator

Cover, Inside Design, and Illustrations by Linda Olliver

PRINTING AND PRODUCTION Catherine Lorrain, Director

NATIONAL SCIENCE TEACHERS ASSOCIATION
Gerald F. Wheeler, Executive Director
David Beacom, Publisher

1840 Wilson Blvd., Arlington, VA 22201
www.nsta.org/store
For customer service inquiries, please call 800-277-5300.

NSTA is committed to publishing material that promotes the best in inquiry-based science education. However, conditions of actual use may vary and the safety procedures and practices described in this book are intended to serve only as a guide. Additional precautionary measures may be required. NSTA and the authors do not warrant or represent that the procedures and practices in this book meet any safety code or standard of federal, state, or local regulations. NSTA and the authors disclaim any liability for personal injury or damage to property arising out of or relating to the use of this book, including any of the recommendations, instructions, or materials contained therein.

Library of Congress Cataloging-in-Publication Data

Keeley, Page.
Uncovering student ideas in science / by Page Keeley, Francis Eberle, and Lynn Farrin.
    v. cm.
Includes bibliographical references and index.
Contents: v. 1. 25 formative assessment probes
ISBN 0-87355-255-5
1. Science--Study and teaching. 2. Educational evaluation. I. Eberle, Francis. II. Farrin, Lynn.
III. Title.
Q181.K248 2005
507'.1--dc22
                            2005018770

The ISBN for *Uncovering Student Ideas in Science, Volume 2* is 978-0-87355-273-8.
eISBN 978-1-933531-73-1

VOL. 2

Uncovering
Student Ideas
in Science

25 More Formative Assessment Probes

Contents

Life Science Assessment Probes

Earth and Space Science Assessment Probes

Foreword

While writing this foreword, I found myself revisiting the 50-odd years of my involvement in science education. I recalled the many ideas, techniques, concepts, and research findings that have passed through my experience and flowed into my teaching repertoire like so much effluent through the filtering rushes in a stream. Some remain vital today and others still cling to the stalks, tried, tested, and found wanting.

I remember so vividly the night of October 4, 1957, when as a nation we were alerted to the beeping of Sputnik as it circled our planet, totally unaware of the influence its presence would have on science education over the next decade. It marked not only the beginning of the space race but the beginning of the rapid and frantic attempts of our nation to "beef up" the science, math, and engineering skills of our students. Science finally had a real place in the school curriculum. The Russians had beaten us to space and we were worried about our future as a nation! The United States responded swiftly with the National Defense Education Act, which allowed teachers like myself to update our content at summer institutes and provided for the development of a different kind of curriculum for school science.

Since then there have been many innovations in our field, including the famed "alphabet soup" curriculum projects of the 1950s and 1960s (e.g., SCIS, SAPA, COPES, Harvard Project Physics) and subsequent curriculum projects such as Insights, GEMS, AIMS, STC, and FOSS.

Then came the advent of the standards decade with Project 2061 and the *Benchmarks for Science Literacy* (AAAS 1993) and the *National Science Education Standards* (NRC 1996). We finally had a guide to what content should be taught and how it should be presented. Many of the states then developed their own versions of the standards, but there was uncertainty about how to use standards on the local level.

In 2005, Page Keeley authored *Science Curriculum Topic Study: Bridging the Gap Between Standards and Practice,* which was the first comprehensive guide to help us bridge the gap between the two sets of national standards, research on student learning, and teaching practice. This was a timely, much-needed book.

Following the development of state standards, each state instituted ways to hold schools accountable for teaching to the standards. For

Foreword

many states, this resulted in "high-stakes" tests, which were enshrined in legislation. Schools gave these tests to students in the spring and received the results sometime during the next school year. The accountability factor was there, but it did little for the teachers who wanted to improve current learning for their students. Many school districts implemented a teaching unit for selected grades entitled "Review for the Test." I thought to myself, "Maybe this really is a good time to retire!"

Many of us believed that teachers needed a way to find out what their students knew, what kind of preconceptions students brought to the classroom, and what teachers could do with this information to improve instruction. Again, Page Keeley and her team from the Maine Mathematics and Science Alliance entered the picture, along with the National Science Teachers Association, with the first volume of *Uncovering Student Ideas in Science: 25 Formative Assessment Probes,* published in 2005. This book focused on helping teachers discern their students' thinking about different science topics. It also helped teachers figure out what to do with this information and where to find help in moving their students to a new and deeper understanding of science concepts.

A workable strategy for formative assessment was now available to the busy teacher. The probes published in the first volume of *Uncovering Student Ideas in Science* were a success, and teachers from all over the country began to find that formative assessment can help them become better teachers. This may indeed have been an example of the "tipping point"

that Malcolm Gladwell (2000) talks about in his book *The Tipping Point: How Little Things Can Make a Big Difference.* I knew it was mine. Finding this kind of innovation is exciting to me because teachers once again can be in charge of classroom instruction. The arrival of a truly inquiry-based focus on science education, coupled with assessment, is what I and so many others have been waiting for.

Well, doesn't a successful book deserve a sequel? Here it is, with 25 new probes and accompanying teacher guides. This is the kind of innovation that is enough to keep an old dog like me barking out there in the field for a few more years. Woof!

Dr. Richard Konicek-Moran
Professor Emeritus
University of Massachusetts, Amherst

References

American Association for the Advancement of Science (AAAS). 1993. *Benchmarks for science literacy.* New York: Oxford University Press.

Gladwell, M. 2000. *The tipping point: How little things can make a big difference.* Boston: Back Bay Books.

Keeley, P. 2005. *Science curriculum topic study: Bridging the gap between standards and practice.* Thousand Oaks, CA: Corwin Press.

Keeley, P., F. Eberle, and L. Farrin. 2005. *Uncovering student ideas in science: 25 formative assessment probes.* Vol. 1. Arlington, VA: NSTA Press.

National Research Council (NRC). 1996. *National science education standards.* Washington, DC: National Academy Press.

Preface

Overview

Since the release of the first volume of *Uncovering Student Ideas in Science: 25 Formative Assessment Probes* (Keeley, Eberle, and Farrin 2005), science educators have shown widespread interest in using formative assessment probes to identify the variety of ideas students bring to their learning and to design instruction based on these preconceptions. This shift from an overemphasis on summative assessment at the end of instruction to a balanced system of formative and summative assessment that happens before, throughout, and at the end of instruction has occurred at the practitioner, researcher, and even policy levels. To understand the reasons for this shift, it will help to briefly review the evolution of formative assessment.

As with the acceptance of new science knowledge and theories, so the emergence and building of new ideas can result in new understandings. Typically a new idea in science is not discovered without previous study and research that has collected a body of evidence in support of the new idea. As the evidence begins to mount and become overwhelming, a point is reached in which the idea becomes accepted and "discovered," often resulting in a new paradigm (Kuhn 1962). The recent research and discoveries that support formative assessment have come about in a similar fashion, causing a paradigm shift in assessment beliefs and practices. While we cannot list everyone who has contributed to the recent "revolution" in accepting formative assessment as a powerful classroom strategy, we would like to acknowledge several of the researchers, assessment specialists, science educators, and practitioners who have sparked our interest, expanded our knowledge base, and helped shape the ideas we include in this book.

Some of the early pioneers in examining students' ideas in science during the 1980s were Rosalind Driver, Edith Guesne, Andrée Tiberghien, Wynne Harlen, Roger Osborne, and Peter Freyberg. They were instrumental in raising science educators' awareness of the use of students' ideas in science as a starting point for effective instruction. In the 1990s Audrey Champagne, Bonnie Shapiro, Lillian McDermott, and Jim Minstrel further articulated the different purposes and kinds of diagnostic, formative, and summative informa-

Preface

tion that science teachers can gain through assessment. Philip Sadler and Matthew Schneps brought us video examples through the *Private Universe Project* (Harvard-Smithsonian Center for Astrophysics 1995), which showed the wide range of ideas students and adults hold, even after instruction. In the same video series, Dick Konicek helped us understand the power of constructivist teaching that takes into account students' ideas. The seminal work of the American Association for the Advancement of Science (AAAS) Project 2061 made explicit links between student ideas and K–12 student learning goals. The summaries of the cognitive research on students' learning of particular concepts and ideas in science appeared in Chapter 15 of the *Benchmarks for Science Literacy* (AAAS 1993), tying research to a clear set of K–12 learning goals. The standards or benchmarks for K–12 student learning in science were now supported by a body of research.

In the late 1990s and continuing to the present, many books and articles about assessment by researchers and practitioners reached educators. Often, however, these publications failed to spell out how formative assessment can be used to improve science instruction and learning. These books were written for a broad audience of practitioners across content areas and lacked connections to the specific nuances of science as a discipline. Research from the cognitive sciences that raised the profile of formative assessment in the science classroom began to reach practitioners with the publication of *How People Learn: Brain, Mind, Experi-*

ence, and School (Bransford, Brown, and Cocking 1999) and *How Students Learn: Science in the Classroom* (Donovan and Bransford 2005). These publications helped us understand how to create and use an assessment-centered environment that acknowledged the importance of starting with students' preconceptions, teaching for transfer, and the role of metacognition.

While new ideas about assessment were emerging in the United States, significant findings in regard to formative assessment were being implemented and disseminated in the United Kingdom. There, researchers and practitioners published resources for teachers that included a variety of science assessment strategies designed to elicit students' ideas and spark inquiries; these ideas and inquiries could lead students to construct new understandings that resolve the dissonance between their preconceptions and scientific explanations (Naylor and Keogh 2000).

The metastudy of formative assessment by Black and Wiliam (1998) crystallized the purposes and effectiveness of formative assessment in instruction as "assessment for learning" rather than "assessment of learning." Black and Wiliam provided evidence for educators that formative assessment is a powerful instructional strategy and includes a variety of forms and purposes. They described how assessment is purposefully used to guide and inform instruction, not to just note in some formal or informal fashion what students are thinking. They further articulated how formative assessment plays out in the science classroom (Black and Harrison 2004).

In 2003 the Maine Mathematics and Science Alliance received a National Science Foundation grant to develop a set of materials to help K–16 educators link national standards and research on student learning to classroom practice. The resulting publication, *Science Curriculum Topic Study: Bridging the Gap Between Standards and Practice* (Keeley 2005), describes the process used to develop the probes in this book. This process links the concepts and ideas from national and state standards to the research on student misconceptions. The information is then used to develop formative assessment probes that reveal the range of ideas noted in the research as well as unique ideas some individual students may hold. The process was applied to develop the first set of 25 probes in Volume 1 of *Uncovering Student Ideas in Science* (Keeley, Eberle, and Farrin 2005) and has been used extensively in professional development to help teachers develop their own probes. Together, these two publications and this new book comprise a powerful set of tools to enhance and extend K–12 science teachers' use of formative assessment.

Collectively, these evolving contributions by researchers, assessment specialists, science education specialists, and practitioners have informed our development of the assessment probes for the *Uncovering Student Ideas in Science* series. It is our hope that the books in this series will support an idea-centered classroom in which teachers use the probes in conjunction with a variety of instructional techniques and questioning strategies. Such instructional practice can make students' thinking and learning visible for the purpose of guiding both students and teachers through the learning process.

Formative assessment is a key feature of classrooms where successful teaching and learning are taking place. The environment of an assessment- and idea-centered classroom is one in which students feel safe to express their ideas, know their ideas are important regardless of whether they are right or wrong, engage in deep thinking and reflection, and have opportunities to test their ideas to revise and improve their thinking. We hope this book can support such an environment.

Next Steps

Uncovering Student Ideas in Science is planned as a series of formative assessment probe books, each volume describing a new application and providing 25 new probes. Volume 1 provided an overview of formative assessment and formative assessment probes. This volume (Volume 2) focuses on ways to use formative assessment to teach for conceptual change. In the third volume of *Uncovering Student Ideas in Science,* we will describe ways teachers can individually use the probes for their professional development as well as ways to develop professional learning communities that engage teachers in examining student work and thinking.

References

American Association for the Advancement of Science (AAAS). 1993. *Benchmarks for science literacy.* New York: Oxford University Press.

Preface

Black, B., and C. Harrison. 2004. *Science inside the black box: Assessment for learning in the science classroom.* London: nferNelson.

Black, P., and D. Wiliam. 1998. Inside the black box. *Phi Delta Kappan* 80 (2): 139–148.

Bransford, J. D., A. L. Brown, and R. R. Cocking. 1999. *How people learn: Brain, mind, experience, and school.* Washington, DC: National Academy Press.

Donovan, M. S., and J. D. Bransford, eds. 2005. *How students learn: Science in the classroom.* Washington, DC: National Academies Press.

Harvard-Smithsonian Center for Astrophysics. 1995. *Private Universe Project.* [Teacher workshop series.] Videotape. Burlington, VT: Annenberg/CPB Math and Science Collection.

Keeley, P., and C. Rose. 2006. *Mathematics curriculum topic study: Bridging the gap between standards and practice.* Thousand Oaks, CA: Corwin Press.

Keeley, P., F. Eberle, and L. Farrin. 2005. *Uncovering student ideas in science: 25 formative assessment probes.* Vol. 1. Arlington, VA: NSTA Press.

Kuhn, T. 1962. *The structure of scientific revolutions.* Chicago, IL: University of Chicago Press.

Naylor, S., and B. Keogh. 2000. *Concept cartoons in science education.* Cheshire, UK: Millgate House.

Acknowledgments

The assessment probes in this book have been extensively piloted and field-tested by the Maine Mathematics and Science Alliance with hundreds of students in northern New England. We would like to thank the many teachers who willingly piloted and field-tested items, shared student data, and contributed ideas for developing new probes. We apologize if we overlooked anyone. In particular we would like to thank the following individuals for their contributions and support of this project:

Lori Agan, ME; Julie Barry, ME; Mary Belisle, ME; Anita Bernhardt, ME; Jodi Berry, ME; Andrew Bosworth, ME; Tracy Bricchi, NH; Ruth Broene, ME; Nancy Chesley, ME; Gay Mary Craig, VT; Elizabeth Crosby, ME; Morgan Cuthbertson, ME; Linda D'Apolito, ME; Laurette Darling, ME; Tad Dippel, VT; Patricia Dodge, VT; Mary Dunn, ME; Mary Evans, ME; Sandra Ferland, ME; Barbara Fortier, ME; Jan Gauger, ME; Jill Gilman, ME; Anne B. Guerriero, NH; Libby Gurnee, ME; Douglas Hodum, ME; Linda Hoffman, ME; Jim Irish, ME; Lisa Jeralds, ME; Leslie Johnson, ME; Vincent Johnson, ME; Barbara Keene, ME; Shawn Kimball, ME; Kathleen King, ME; Susan Kistenmacher, ME; Mary Anne Knowles, ME; Linda Kutyz, ME; Peggy LaBrosse, NH; Cindy Langdon, ME; Gary LaShure, VT; Joanna Leary, ME; Lee Leoni, ME; Anne Macdonald, ME; Cheryl Marvinney, ME; Patty Mendelson, ME, Kris Moniz, ME; Wendy Moore, VT; Margo Murphy, ME; Andrew Njaa, ME; Laurie Olmsted, ME; Dr. Lois K. Ongley, ME; Jack O'Reilly, NH; Cindy Pellerin, ME; Ingrid Porter, ME; Linda Prescott, ME; Laura Reed, VT; Ruth Scheibenpflug, ME; Cyrene Slegona, ME; Jane Voth-Palisi, NH; R. David White, ME; and Mary Whitten, ME.

We also thank the following individuals for their reviews of this book as well as for providing feedback on individual probes: Al Colburn, Jo Dodds, Chad Dorsey, Carolyn Landel, Cheryl and Bob Marvinney, Michael Tinnesand, and Anne Tweed.

About the Authors

Page Keeley is the senior program director for science at the Maine Mathematics and Science Alliance (MMSA), where her work involves supporting leadership, professional development design, mentoring, formative assessment, and development of tools and resources to support science teachers. She is the principal investigator/project director of three National Science Foundation projects: the Northern New England CoMentoring Network (NNECN), Curriculum Topic Study (CTS)—A Systematic Approach to Utilizing National Standards and Cognitive Research, and PRISMS (Phenomena and Representations for the Instruction of Science in Middle School). She is also the primary author of four other books: *Science Curriculum Topic Study: Bridging the Gap Between Standards and Practice* (Keeley 2005); *Uncovering Student Ideas in Science: 25 Formative Assessment Probes,* Vol. 1 (Keeley, Eberle, and Farrin 2005); *Mathematics Curriculum Study: Bridging the Gap Between Standards and Practice* (Keeley and Rose 2006), and *Science Formative Assessment: 75 Practical Strategies for Linking Assessment, Instruction, and Learning* (Keeley, Forthcoming). Page serves on several national advisory boards and consults with schools and organizations nationally. Be-

fore working at the MMSA, she taught middle and high school science for 15 years. During that time she served as an NSTA District II director, a two-term member of the NSTA Executive Board, and president of her state science teachers association. She has taught as an adjunct instructor at the University of Maine. Page received the Presidential Award for Excellence in Secondary Science Teaching in 1992 and the Milken Family Foundation National Distinguished Educator Award in 1993. Before beginning her teaching career, she worked as a research assistant in immunology at the Jackson Laboratory of Mammalian Genetics in Bar Harbor, Maine. She received her BS in life sciences from the University of New Hampshire and her MEd in science education from the University of Maine.

Dr. Francis Eberle is the executive director of the MMSA. He has been a principal investigator or co-principal investigator on six National Science Foundation K–16 STEM (science, technology, engineering, and mathematics) education projects, including Maine's Statewide Systemic Initiative. He has led several school-based science and mathematics projects, including two state mathematics and science partnership grants. Currently he is focusing his work on the use of data at both the formative and summative levels and understanding teacher beliefs to inform improvements in science teaching. Francis has written several articles on science instruction, formative assessment, attributes of effective classrooms, and science standards, and he is a coauthor of

Preface

the book *Uncovering Student Ideas in Science: 25 Formative Assessment Probes,* Vol. 1. Currently he serves as president of the National Association of State Science and Mathematics Coalitions (NASSMC). Before coming to the MMSA in 1993, Francis taught middle and high school science, served as an adjunct instructor at the University of Southern Maine, and founded and directed a nonprofit organization, the STAR Science Center. He holds a BS in science education from Boston University, an MS in educational psychology from the University of Connecticut, and a PhD from Lesley University in science educational studies.

Joyce Tugel is a science specialist at the MMSA. Her work focuses on teacher leadership, mentoring and new teacher support, and science professional development. She is currently the project director on a state Mathematics and Science Partnership Project, Science Content, Conceptual Change, and Collaboration (SC4). Before coming to the MMSA in 2005, Joyce was the science professional development specialist at the TERC Eisenhower Regional Alliance for 5 years and taught high school chemistry and physical science for 10 years in southern Maine. During that time, she received the Presidential Award for Science Teaching in 1998, the Milken Family Foundation National Distinguished Educator Award in 1999, and the New England Institute of Chemists Secondary Teaching Award in 1999. Joyce has served as an NSTA District II director and NSTA's Professional Development Division director. Before beginning her teaching career, she worked in research as an environmental biogeochemist at the University of New Hampshire. She received her BS and MS degrees in microbiology from the University of New Hampshire.

Introduction

Assessment for learning is any assessment for which the first priority in its design and practice is to serve the purpose of promoting pupils' learning. It thus differs from assessment designed primarily to serve the purposes of accountability, or of ranking, or of certifying competence.

—Paul Black and Christine Harrison,
Science Inside the Black Box: Assessment for Learning in the Science Classroom

Probes as Assessments for Learning

Imagine a second-grade classroom where students are having a science talk about rocks. The teacher uses the "Is It a Rock?" probe to make a list of rock-like items of different shapes and sizes for the students to consider. She shows them pictures or actual samples of the items on the list such as boulders, pebbles, and a grain of sand. Students are asked to share their ideas about which items on the list they would call rocks and why. Some students argue that stones and pebbles are not rocks because they are too small. Other students argue that a rock has to be rough and jagged. If it is smooth, it is no longer a rock but, instead, a stone. Still other students argue that size and shape do not matter. If it is made up of rock, then it is a rock. The entire time the teacher is orchestrating the science talk, she is listening carefully to the students' ideas and the reasoning they use to support their concept of a rock. She is using their ideas to think about the experiences she might provide next to help the students come up with a common "rule" that would determine whether something is a rock, leading toward a generalized concept of *rock,*

Introduction

regardless of size or shape.

In a fourth-grade classroom, students are sharing ideas about things seeds need to grow. The class is undecided as to whether seeds must have darkness to sprout. After giving students the probe "Needs of Seeds," the teacher shares with the class several student responses to the probe that stated that seeds must have darkness. The responses describe how seeds need to be in the soil, away from sunlight, in order to sprout and grow into a seedling. When the students are asked why they thought the seeds need to be kept away from light, they draw on their prior experiences planting seeds in a garden by covering them with soil. A few students think the soil is needed to provide moisture, not to keep out the light. Most students think soil is necessary and that seeds would not sprout without soil. The teacher realizes that students need to have experiences germinating seeds under various conditions in order to confront them with some of their misconceptions. When asked how they could test their ideas, the students come up with several experiments that would support the ideas they have about what seeds need to sprout. The teacher provides the materials for students to test their ideas, after which they will come back together to share results and decide if their findings changed any of their ideas about needs of seeds.

In a middle school classroom, students individually complete the probe "Comparing Cubes" and then form small groups to discuss their ideas. They listen carefully to each others' ideas and try to reconcile their different ideas about how size affects the properties of objects made out of the same materials. The teacher observes several students trying to persuade their classmates to consider their idea and looks for evidence of conceptual change as students share their thinking.

In the first week of a high school chemistry class, students are asked to complete the "What's in the Bubbles?" probe. The teacher collects the student work and is surprised to find that most of the students thought the bubbles were filled with air. Although the students want to know the right answer, she assures them that they will discover it for themselves during their unit on gases and changes in state. She carefully plans her instruction so that students will encounter the idea in the context of their lab activities. After students complete the activities, she returns the probe that they completed several weeks before and gives them an opportunity to revise or expand on their previous answers. Several students are surprised to see that they once thought that bubbles of boiling water were filled with air or contained nothing. In their reflections they remark how their ideas changed based on the activities that proved there was some form of water, which they now call water vapor, in the bubbles.

What do all four of these examples have in common? Each of these teachers is using formative assessment probes as assessments *for* learning rather than assessments *of* learning. The probes enhance the interactions between students and between students and teachers by providing an engaging context to elicit the preconceptions students bring to their learning and a focus for students to examine and talk through their ideas. While the probes promote student thinking, they also provide valuable information to the teacher to use for making instructional decisions. The teacher uses the students' ideas to design instruction that will build a bridge between students' preconceptions and the scientific knowledge they need to understand and explain everyday objects, processes, and phenomena. Building this bridge involves the process of conceptual change.

The conceptual change model (CCM) of instruction was first described by Posner and colleagues in 1982. In instruction based on this model, teachers use elicitation strategies such as the probes in this book to draw out students' ideas and make students' thinking known to both the student and the teacher. Once ideas are exposed, instruction is designed so that students will have an opportunity to be confronted with their ideas. Several of the probes in this book provide an opportunity for students to test their ideas and discover that their observations did not match their predictions. The probes can also be used to pose questions for class discussion, which provides another opportunity for students to be confronted with ideas that may not match their thinking. When students experience dissonance between the ideas and beliefs they hold and what they are experiencing, they need to resolve the dissonance and accommodate new ideas that may provide better explanations than the ones they previously held. Throughout the CCM model of instruction, teachers are continuously monitoring student learning and adjusting instruction so that students can progress through the

changes that will eventually lead to a correct scientific conception and enduring understandings.

Linking Probes, Teaching, and Learning

This interplay between teaching and learning, based on uncovering students' ideas, is the essence of formative assessment using the probes provided in this book. Formative assessment has a specific purpose distinguishing it from summative assessments such as quizzes, tests, and performance tasks. The probes in this book are designed to be seamlessly embedded into classroom instruction. They are used to assess *before, during, and throughout* the teaching and learning process, rather than at an end point after instruction. Their primary purpose is to promote student thinking and improve opportunities to learn by providing valuable feedback to the teacher, rather than the summative purposes of measuring and reporting student achievement. They are so inextricably linked to teaching that it is often difficult to determine whether a probe is used as an instructional or assessment strategy. The probes in this book offer stimuli that enable learners to interact with assessment in a variety of ways—through writing, drawing, speaking, listening, and designing and carrying out investigations. Unlike other types of assessments, they are an unobtrusive part of teaching. The probes and their accompanying teacher notes can be used

- to uncover students' understanding about a science topic prior to and throughout instruction, informing teachers in short-

Introduction

term lesson planning, long-term unit development, and even longer-term review and modifications for units taught again the next year;

- as a teaching and learning activity to engage students, spark inquiry, and stimulate thinking;
- to help teachers self-assess and analyze their own teaching by examining how well students are progressing toward a conceptual understanding of scientific ideas;
- to determine whether students need to experience ideas in new and varied contexts;
- to motivate and enhance student learning by helping students recognize that their ideas are valued and taken into account to design instruction that meets their needs;
- to provide feedback to both teachers and students so that as teachers and students respond to each others' ideas, both teaching and learning are supported;
- to promote safe, rich discourse in the classroom that recognizes that everyone's ideas are important regardless of whether they are right or wrong;
- to provide a familiar context for English-language learners to develop language skills that support science discourse and science learning;
- to differentiate instruction for different groups of students that targets their preconceptions and takes into account their unique experiences and backgrounds that shape their thinking; and
- to stimulate and motivate teachers to try

out new instructional practices that may be more effective in producing conceptual change.

Research Supporting the Use of Probes

How People Learn: Brain, Mind, Experience, and School (Bransford, Brown, and Cocking 1999) presents research applicable to how students learn science that has implications for what we teach, how we teach it, and how we assess it. Two of the findings from this report strongly support the use of formative assessment probes:

1. *"If [the students'] initial understanding is not engaged, they may fail to grasp new concepts and information presented in the classroom, or they may learn them for purposes of a test but revert to their preconceptions"* (p. 14). The primary purpose of the probes is to elicit the ideas students have before engaging in a learning experience. By knowing in advance what these ideas are, teachers can design targeted instruction and monitor students' learning throughout the sequence of instruction. This differs from the traditional use of assessments given at the end of instruction, which may reveal students' conceptual difficulties after a unit of instruction has ended and the opportunity to design targeted instruction is gone.

2. *"A 'metacognitive' approach to instruction can help students learn to take control of their own learning by defining*

learning goals and monitoring their progress in achieving them" (p. 18). Each of the probes is designed to target a specific learning goal, as described in the teacher notes that accompany each probe. By linking a probe to a learning goal and encouraging students to make their own thinking explicit, probes promote metacognition. They help students think about their own ideas and whether or not they are learning.

How People Learn has changed our view of how classroom environments should be designed. The following aspects of classroom environments relate directly to classroom climates and cultures where formative assessment probes are used in instruction:

- *Learner-Centered Environment* (p. 23). In a learner-centered classroom where teachers are using probes before and throughout instruction, teachers pay careful attention to the progress of each student and know at all times where their students are in their thinking and learning.
- *Knowledge-Centered Environment* (p. 24). In a knowledge-centered environment, teachers know the goals for learning, the key concepts and ideas that make up the goals, the prerequisites on which prior and later understandings are built, the types of experiences that support conceptual learning, and the assessments that will provide information about student learning. All of these considerations are described in the teacher notes that accompany each

probe. These goals, key concepts and ideas, and prerequisite learnings can be made explicit to students as well.

- *Assessment-Centered Environment* (p. 24). Ongoing assessments, such as the probes, are designed to make students' thinking visible to both teachers and students. These assessments are learner-friendly and provide students with opportunities to revise and improve their thinking, help students see their own progress, and help teachers identify problem areas to focus on.
- *Community-Centered Environment* (p. 25). Probes are used to promote social interactions around learning ideas in science. Student learning is supported in classrooms where social discourse is encouraged and norms such as accepting that it is all right not to know the right answer, feeling a sense of safety in academic risk-taking, and encouraging revision of ideas and reflection are supported. Classrooms where students feel part of an intellectual learning community that takes ownership of students' ideas are places where students and teachers thrive.

Taking Into Account Students' Ideas

Teaching science can be difficult because students have already formed ideas about the natural world and how it works before coming to the classroom. One of the challenges teachers face is how to assess student's preconceptions before introducing new topics, so that appropriate learning experiences can be provided.

Introduction

This challenge highlights why the use of formative assessment probes is such an important instructional strategy. In situations where students' views differ from the scientific view, it is very difficult to change conceptions if teachers are unaware of students' ideas before and throughout instruction. In addition, it is not enough to identify and diagnose students' misconceptions. Teachers need to employ deliberate strategies to work with students' ideas and design opportunities for students to construct new understandings that will lead to conceptual change. Much of "school science" is contrary to students' everyday experience and intuition; hence, students' ideas may be highly resistant to change, even after several instructional experiences (Driver et al. 1994). Many teachers do not recognize this unless they continuously check for that understanding on a regular basis and engage students in identifying and challenging their own thinking.

The practical idea of using formative assessment probes is a first step in incorporating student thinking into instruction. Identifying students' ideas before beginning a lesson or unit can be quite informative, but it is just the first step in improving student learning. Once students' ideas have surfaced, making informed determinations about what can be done to move students toward the scientific view is the next step. Driver et al. (1994, p. 10) suggest that the first step might be to consider the nature of the differences between students' thinking and the scientific view. As a result of probing for students' preconceptions, teachers might do the following:

- *Develop existing ideas.* Students' ideas may be close to the scientific idea, but not fully developed. If so, instruction can be focused toward developing a deeper understanding of the scientific idea. For example, teachers might use the probe "Floating Logs" and discover that students understand that the logs will float in the same way because the shape and material are the same. However, they may not be able to explain it in terms of the unchanging characteristic property of density. Their notion of sameness can be further extended in developing the concept of density as a constant proportional relationship between mass and volume.

- *Differentiate among existing ideas.* Students may have similar ideas about objects or phenomena that cause them to overgeneralize or categorize too broadly. If so, instruction can be focused on helping students differentiate ideas. For example, teachers might use the "Is It a Rock? (Version 2)" probe to find out if students think any rock-like material, whether geologic in origin or human made, is considered to be rock. Teachers can use this information to help students differentiate between rocks formed by natural, long-term geologic processes and rock-like materials that were human made and formed in a short time.

- *Integrate existing ideas.* Students may form separate ideas, often in different contexts. For example, teachers might use the probe "Is It Food for Plants?" to help students integrate the idea of what food is, in a biological context, with the needs plants have in carrying out their life processes. They may have previously learned about food in the context of nutrition and studied the life processes of plants and animals but may have failed to integrate the idea that plants not only make their own food but use it in ways similar to animals' use of food.

- *Changing existing ideas.* Students often have misconceptions about objects, materials, or processes. For example, teachers might use the probe "Emmy's Moon and Stars" to find out that students have misconceptions about where stars are located. The information is used to design instruction that addresses the relative scale of vast distances between visible objects in the night sky and that will help students change their existing ideas and accept the scientific view.

- *Introduce new ideas.* Students should have opportunities to demonstrate when they are ready for a new idea or ready to build more sophisticated understandings of an existing idea. For example, teachers might use the "Baby Mice" probe to learn that their students have the correct idea that both parents contribute genetic material that determines an offspring's traits. Teachers can build on this simple idea by introducing the more sophisticated details about the mechanism for inheritance.

Suggestions for Embedding Probes in Instruction

The probes developed for *Uncovering Student Ideas in Science* can be embedded into teaching

in a variety of ways, depending on the phase of instruction and the instructional strategy they are used with. The following is a list of suggestions for using the probes to engage students in investigating new ideas and phenomena, elicit students' ideas, help them be more aware of their thinking, and construct new ideas through discourse with their peers.

Use the probes to initiate scientific inquiry. Several of the probes are designed so that students can make predictions and test their ideas (e.g., "Boiling Time and Temperature"). The familiar context of the probe is designed to relate to students' everyday experiences and engage them in wanting to know whether their ideas can be confirmed through investigation. Ask students to share previous experiences, knowledge, or ideas they based their predictions on and design an investigation to test their predictions.

Make students' thinking explicit during scientific inquiry. Use the probes to draw out students' thinking before and throughout an inquiry. Encourage students to use evidence to construct their explanations both before and during the investigation, particularly when the evidence does not match their preconceptions. The dissonance that results when data do not support their original thinking is the pivotal point in helping students reject their former ideas in order to accommodate new ones.

Create a culture of ideas, not answers. Use the probes to encourage students to share their

Introduction

ideas, regardless of whether they are right or wrong. Students have been raised in a school culture where they are expected to give the "right answer." Thus they hesitate to share their own ideas when they think they may be "wrong." Hold off on telling students whether they are right or wrong and give them an opportunity to work through the ideas, weighing various viewpoints and evidence, until they are ready to construct an understanding. The emphasis on testing and revising one's conceptual model should take precedence over getting the right answer. Getting all ideas out on the table first may be frustrating and take longer, but in the long run it will develop deeper, more enduring understanding and students will be less apt to revert back to their previous conceptions after the unit of instruction ends.

Develop a discourse community. Use the probes as a way to get students to talk to one another, rather than the typical back-and-forth "ping-pong" discussion between teacher and students that is typical of a non-inquiry-based classroom. Strategies such as "think-pair-share"—where students first think about their own ideas (perhaps even writing them down on the probe sheet), then pair up with another student to discuss their ideas, and finally share their ideas either in a small group or with the whole class—provide opportunities for students to be more involved in discussion about ideas and take an active role in developing a community of learners. Use the probes for "volleyball," not "ping-pong." In other words, the discussion should move back and forth between several

students and groups of students before it comes back to the teacher for a new question or comment. The type of discourse is as important as the patterns of "science talk." The discourse must value alternative points of view and not just the authoritative view of the teacher.

Encourage students to take risks. Create a climate where it is acceptable to go out on a limb with an idea in a social context without being put down by the teacher or other students. One way to do this is to use "no hands questioning," where students are randomly called on to share their thinking (Black et al. 2003) and there is an expectation that everyone has ideas worth sharing, no matter how tentative they are. Create norms of collaboration in the classroom so that everyone's ideas are respected and acknowledged.

Encourage students to listen carefully. In a formative assessment classroom, different ideas are shared among pairs of students, small groups, and the whole class. Students need to learn to listen carefully to others' ideas and weigh the evidence before changing their own ideas. They need to learn not to accept a new idea simply because their peers think it is correct. They need to learn how to examine all the ideas, including evidence from investigation and other relevant information sources, before accepting an idea or changing a previously held one. Formative assessment encourages students to think, rather than just accept ideas as they are presented.

Use the probes in a variety of ways. Don't use the probes solely as a written assessment.

Use them in pair, small-group, and class discussions. Vary strategies for sharing responses. For example, students can form groups based on the response they selected to discuss their ideas, then jigsaw with other groups to consider alternative explanations. Responses can be shared anonymously by the teacher, allowing the class to discuss and evaluate different explanations. Encourage students to use drawings or whiteboards to illustrate their thinking in addition to using words. Consider using the probes in interviews with students, which allow the teacher to probe even more deeply. *Science Formative Assessment: 75 Practical Strategies for Linking Assessment, Instruction, and Learning* (Keeley, Forthcoming) provides a wealth of practical techniques for using probes during various stages of teaching and learning.

Encourage continuous reflection. Encourage students to reflect back on their initial ideas about the probe throughout a sequence of lessons in order to note evidence of their own conceptual changes or to identify areas where they are still struggling with an idea. Understanding is an evolving process. It takes time for students to move toward the accepted scientific view, and students need to understand that there are many steps along the way. Being aware of one's thinking (metacognition) and knowing what one's learning goal is will help students be more accountable for their own learning. Revisiting their initial responses to the probe and comparing them with where they are in their current understanding is a powerful metacognitive strategy.

Our ideas and suggestions for using the probes stem from elements of best practice in science teaching, which uses students' ideas as the starting point for instruction, continuously monitors students' progress toward conceptual development and change, and engages students in the teaching and learning process. One might ask, "Well, what does it look like in practice?" The vignette on pages 13–15 provides a middle school example of how a teacher might use the probes to integrate assessment with instruction (see Volume 1 for additional vignettes).

Using the Teacher Notes That Accompany the Probes

The probes by themselves are powerful for eliciting students' ideas. Just as powerful for the teacher are the notes that accompany each probe. Effective science teaching takes into account students' developmental level, grade levels where ideas are taught in the standards, specific ideas in the standards, research on commonly held ideas, effective instructional strategies, and the need for additional assessments. A science teacher uses all of these considerations to adjust instruction and provide learning opportunities that challenge students' existing conceptions, while monitoring students' journeys toward a more scientific understanding. Each of the probes in this book contains detailed teacher notes to help you best decide how, when, and with what grade to use the assessment probe; link the ideas addressed by the probe to related standards; examine research that informed the development of the

Introduction

probe and provides additional insight into students' thinking; consider new instructional strategies to help students learn the ideas; and access additional information to learn more about the topic addressed by the probe. The following sections describe each component of the teacher notes.

Purpose

This section describes the general purpose of the probe. It includes the concept or general topic as well as a description of the specific idea that the probe is intended to elicit. It is important that you be clear about what the probe is going to reveal so that you can decide if the probe fits your intended target.

Related Concepts

Each probe is designed to target one or more related concepts that cut across grade spans. A single concept may be addressed by multiple probes, and you may find it useful to use a cluster of probes to target a concept or specific ideas within a concept. For example, there are several probes that target the concept of density. The concept matrices on pages 18, 92, and 150 can help you identify related probes.

Explanation

A brief scientific explanation accompanies each probe to provide clarification of the scientific content that underlies the probe. The explanations are designed to help you identify what the most scientifically acceptable answers are (sometimes there is not a "right" answer) as well as clarify any misunderstandings you might

have about the content. The explanations are not intended to provide detailed background knowledge on the concept, but they should provide enough information to connect the idea in the probe with the scientific knowledge it is based on. If you need further explanation of the content, the teacher notes list National Science Teachers Association (NSTA) resources, such as the *Stop Faking It! Finally Understanding Science So You Can Teach It* series, that will enhance and extend your understanding of the content.

Curricular and Instructional Considerations

The probes in this book are not limited to one grade level in the way that summative assessments are. What makes them interesting and useful is that they provide insights into the knowledge and thinking that students in your school may have related to a topic as they progressed from one grade level to the next. The assessment of ideas that students may not formally encounter until later in their education may help teachers in the later grades understand where and how preconceptions originate. Some of the probes can be used in grades K–12; others may cross over just a few grade levels. Teachers in two different grade spans (e.g., middle and high school) might decide to use the same probe and come together to discuss their findings. Since the probes do not prescribe a specific grade level for use, you are encouraged to read the curricular and instructional considerations and decide if your students have had sufficient experience to make the probe useful in your instructional context.

This section may also describe how the information gleaned from the probe is useful at a given grade span. For example, the information might be useful for planning instruction when an idea in the probe is a grade-level expectation, or it might be useful in a later grade to find out whether students have sufficient prior knowledge to move on to the next level of understanding. Sometimes the knowledge gained through use of the probe indicates that you might have to back up several grade levels to teach ideas that have not been fully developed in previous grades.

We deliberately chose not to suggest a grade level for each probe. If the probes had been intended to be used for summative purposes, a grade level, aligned with a standard, would have been suggested. However, these probes have a different purpose. Do you want to know about the ideas your students are expected to learn according to your grade-level standards? Are you interested in how preconceived ideas develop and change across multiple grade levels in your school even when they are not formally taught? Are you interested in whether students achieved a scientific understanding of previous grade-level ideas before you introduce higher-level concepts? The descriptions of grade-level considerations in this section can be coupled with the section that lists related ideas in the national standards in order to make the best judgment about grade-level use.

Several of the elementary, middle, and high school grade-level curricular and instructional considerations have an embedded SciLink. NSTA SciLinks staff have identified websites that match the grade level and topic of the probe. By going to the SciLinks website, *www.scilinks.org,* and entering the code number provided, you can access additional resources that relate to the probe.

Administering the Probe

Suggestions are provided for administering the probe to students, including a variety of modifications that may make the probe more useful at certain grade spans. For example, the notes might recommend eliminating certain examples from a list for younger students who may not be familiar with particular words or examples. This section may also include suggestions for demonstrating the probe context with props or ways to elicit the probe responses while students interact within a group.

Related Ideas in *National Science Education Standards* and *Benchmarks for Science Literacy*

This section lists the learning goals stated in the two national documents generally considered the "national standards": *National Science Education Standards* (NRC 1996) and *Benchmarks for Science Literacy* (AAAS 1993). Since the probes are not designed as summative assessments, the learning goals listed are not intended to be considered as alignments but rather as related ideas connected to the probe. Some targeted ideas, such as a student's conception of a plant in "Is It a Plant?" (p. 93) are not explicitly stated as learning goals in the standards but are clearly related to national standards concepts such as characteristics of

Introduction

living things and classification of objects or living things. When the ideas elicited by a probe appear to be a strong match with a national standard's learning goal, these matches are indicated by a star ★ symbol. You may find this information useful in using probes with lessons and instructional materials that are aligned to national standards and used at a specific grade level.

Related Research

Each probe is informed by related research where available. Since the probes were not designed primarily for research purposes, an exhaustive literature search was not conducted as part of the development process. The authors drew on two comprehensive research summaries commonly available to educators: Chapter 15, "The Research Base," in *Benchmarks for Science Literacy* (AAAS 1993) and Rosalind Driver et al.'s *Making Sense of Secondary Science: Research Into Children's Ideas* (1994). Although the research summarized in these resources was conducted in the 1980s and 1990s, the results of these studies are considered timeless and universal. Children in the United States and in other countries hold ideas that have been found to be pervasive across geographic boundaries, despite the societal and cultural contexts that can influence students' thinking.

The descriptions from the research can help you better understand the intent of the probe and the variety of responses your students are likely to demonstrate when they respond to the probe. As you use the probes, you are encouraged to seek new and additional research find-ings. One source of updated research can be found on the Curriculum Topic Study (CTS) website at *www.curriculumtopicstudy.org*. A searchable database on this site links each of the CTS topics to additional research articles and resources.

Suggestions for Instruction and Assessment

After analyzing your students' responses, it is up to you to decide on the student interventions and instructional planning that would work best in your particular curricular and instructional context. We have included suggestions gathered from the wisdom of teachers, from the knowledge base on effective science teaching, and from our own collective experience as former teachers and specialists involved in science education. These are not exhaustive or prescribed lists but rather a listing of suggestions that may help you modify your curriculum or instruction, based on the results of your probe, to help students learn ideas that they may be struggling with. It may be as simple as realizing that you need to provide a variety of contexts, or there may be a specific strategy or activity that you could use with your students.

Learning is a very complex process, and it is unlikely that any single suggestion will help all students learn the science ideas. But that is part of what formative assessment encourages—thinking carefully about the variety of instructional strategies and experiences need-ed to help students learn scientific ideas. As you become more familiar with the ideas your students have and the multifaceted factors that may have contributed to their misunderstandings, you will identify additional strategies that you can use to teach for conceptual change.

Related NSTA Science Store Publications and NSTA Journal Articles

NSTA's journals and books are increasingly targeting the ideas students bring to their learning. We have provided suggestions for additional readings that complement or extend the use of the individual probes and the background information that accompanies them. For example, Bill Robertson's *Stop Faking It!* series of books may be helpful in clarifying content for students (as well as for teachers). An article from one of NSTA's elementary, middle school, or high school journals may provide additional insight into students' misconceptions or provide an example of an effective instructional strategy or activity that can be used to develop understanding of the ideas targeted by a probe. Other resources listed in this section provide a more comprehensive overview of the topic addressed by the probe. To access the Science Store, go to *www.nsta.org* and click on the link; to access journal articles, go to *www.nsta.org* and click on Teacher Resources, then NSTA Publications, then NSTA journals, and then the relevant journal.

Related Curriculum Topic Study Guides

NSTA is a co-publisher of the book *Science Curriculum Topic Study: Bridging the Gap Between Standards and Practice* (Keeley 2005). This book was developed as a professional resource for teachers with funding from the National Science Foundation's Teacher Professional Continuum Program. It provides a set of 147 CTS guides that can be used to learn more about a science topic's content, examine instructional implications, identify specific learning goals and scientific ideas, examine the research on student learning, consider connections to other topics, examine the coherency of ideas that build over time, and link understandings to state and district standards. The CTS guides use national standards and research in a systematic process that deepens teachers' understanding of the topics they teach.

The probes in this book were developed using the CTS guides and the assessment tools and processes described in Chapter 4 of the CTS book. The CTS guides that were used to inform the development of each of the probes are listed in the teacher notes that follow each probe. Teachers who wish to delve more deeply into the standards and research-based findings that informed the development of the probe may use the CTS guides for further information about teaching and learning connected to the ideas in the probe.

References

References are provided for the standards and research findings cited in the teacher notes.

Vignette on Teaching Density

Before starting my unit on density, I talked with an elementary teacher in my district to find out what kinds of things students had experienced prior to middle school that my density unit could build on. She told me that

Introduction

one of the kits used by all teachers at her grade level includes an activity where students predict and test whether objects will float or sink. This seemed like a great place for me to start with my eighth-grade students. By bringing back a vivid experience from earlier grades, I would be able to build on their existing foundation of knowledge.

When my students arrived in class the next day, I placed a 20 cm length of a wooden dowel in a small tank of water. As it floated, I displayed a piece of another wooden dowel that was made out of the same material, whose length and diameter were double the size of the dowel floating in the tank. I distributed the probe "Floating Logs," and students commented that the dowel in the tank of water was like the floating log portrayed on the sheet of paper. I asked them to envision what they thought would happen if I were to place the larger dowel in the water. Would it float in a similar way or would it float differently? I asked them to respond to the probe, explaining their thinking. I assured them that I was not looking to see if their answer was right or wrong, but I wanted to know what they were really thinking would happen—and, most importantly, why they thought it would happen.

The room became very quiet, and students began busily writing. I was surprised at how much time they took to describe their reasons, because many of them have a difficult time with the essay questions on my tests. In a few minutes, I noticed students were quietly talking with their neighbors, pointing to the dowels at the front of the room. I heard them using terms such as *mass, length, volume,* and *density* but noticed that some of them were confusing extensive properties such as size with the intensive property of density. This seemed like a good time for some "science talk."

We placed our chairs in a circle. I asked them to fold their papers in half and pass them back and forth across the circle multiple times until I said "stop," so other students could not identify the paper in their possession. They were asked to respect the student whose paper they had by not revealing the student's name unless that student spoke up to acknowledge the ideas on the paper and to add to them if needed. I asked for a show of hands: Who had a paper with response A? Response B? Response C? I was surprised by the mixed responses! I asked students to share some of the reasons that were on the paper they had "adopted." Some responses said that if you have more of something, it's more dense. Others said heavier things sink. As students continued to share the reasoning, they started questioning each other. I heard many "What would happen if ..." kinds of questions, and I realized this was a golden opportunity to let students take ownership of their learning.

We brainstormed a list of testable questions, and the next day each team chose a variable to investigate that would help the class find out if the size of an object affects the way in which it floats. Although it was agreed that each team would keep the kind of material they used during their investigation constant, we didn't all have to use the same material. It was interesting to see the variety of materials and ways students chose to quantify differences in size. Some measured volume by water displacement or by using mathematical calculations, and others measured mass. As I circulated around the room, I found that some students were changing their original ideas, based on the new evidence that was gathered. Others were somewhat reluctant to change their ideas and engaged in further discussion to try and figure out why their results were not matching their original ideas. As we gathered for an investigation wrap-up, the class results provided evidence that the amount of material, whether documented by mass or volume, does not affect the way the material floats.

As the students' conceptual understanding of density grew, I began to bring in the scientific terminology. Subsequent lessons connected the proportional relationship of mass and volume to this characteristic property, and I was now ready to introduce the symbolic representations using a mathematical equation.

As I brought the density unit to a close, I asked students to conduct a 10-minute "quiet write":

Re-read your original "Floating Logs" explanation. Has any of your thinking changed? If so, what are you now thinking, and how do you know this? What "rule" or reasoning are you using to explain your new thoughts?

I brought their responses home to read that night and was struck by the power of written reflection. Students were able to use some of our classroom activities to formulate their new "rules," and I could see how their thoughts were influenced by their experiences. "Floating Logs" allowed my students to become aware of their own ideas, and through testing and analyzing they were able to resolve some of the conflict between their initial ideas and the scientific concept.

Concept Matrices and Probe Set

The remainder of this book contains a set of 25 probes that you can use with your students along with accompanying teacher notes on each of the probes. The concept matrices (pp. 18, 92, and 150) indicate the concepts most related to each probe and can be used to select probes that match your instructional context. In this volume we focus on properties of matter, particulate matter, energy, plants, genetics, cells, adaptation, geology, the day/night cycle, and sky objects. Volume 3 will include additional topics in life, Earth, space, and physical science.

References

American Association for the Advancement of Science (AAAS). 1993. *Benchmarks for science literacy.* New York: Oxford University Press.

Black, P., and C. Harrison. 2004. *Science inside the black box: Assessment for learning in the science classroom.* London: nferNelson Publishing Company.

Black, P., C. Harrison, C. Lee, B. Marshall, and D. Wiliam. 2003. *Assessment for learning: Putting it into practice.* Berkshire, England: Open University Press.

Bransford, J. D., A. L. Brown, and R. R. Cocking. 1999. *How people learn: Brain, mind, experience,*

Introduction

and school. Washington, DC: National Academy Press.

Driver, R., A. Squires, P. Rushworth, and V. Wood-Robinson. 1994. *Making sense of secondary science: Research into children's ideas.* London: RoutledgeFalmer.

Keeley, P. 2005. *Science curriculum topic study: Bridging the gap between standards and practice.* Thousand Oaks, CA: Corwin Press.

Keeley, P. Forthcoming. *Science formative assessment: 75 strategies for linking assessment, instruction, and learning.* Thousand Oaks, CA: Corwin Press.

National Research Council (NRC). 1996. *National science education standards.* Washington, DC: National Academy Press.

Posner, G., K. Strike, P. Hewson, and W. Gertzog. 1982. Accommodation of a scientific conception: Toward a theory of conceptual change. *Science Education* 66: 211–227.

Stepans, J. 2003. *Targeting students' science misconceptions: Physical science concepts using the conceptual change model.* Tampa, FL: Idea Factory.

Physical Science Assessment Probes

Probes

Core Science Concepts	Properties of Matter							Particulate Matter		Energy	
	Comparing Cubes	Floating Logs	Floating High and Low	Solids and Holes	Turning the Dial	Boiling Time and Temperature	Freezing Ice	What's in the Bubbles?	Chemical Bonds	Ice-Cold Lemonade	Mixing Water
Atoms or Molecules	✓							✓	✓		
Boiling and Boiling point					✓	✓		✓			
Buoyancy			✓								
Change in State					✓	✓		✓			
Characteristic Properties	✓	✓	✓	✓	✓	✓	✓				
Chemical Bonds									✓		
Conduction										✓	✓
Density	✓	✓	✓	✓							
Energy					✓	✓	✓	✓		✓	✓
Energy Transfer										✓	✓
Extensive Properties of Matter	✓										
Freezing Point							✓				
Heat					✓	✓				✓	✓
Intensive Properties of Matter	✓	✓	✓	✓	✓	✓	✓				
Mass	✓										
Melting Point	✓										
Sinking and Floating	✓	✓	✓	✓							
Temperature					✓	✓	✓				✓
Weight	✓										

ComparingCubes

Sofia has two solid cubes made of the same material. One cube is very large, and the other cube is very small. Put an X next to all the statements you think are true about the two cubes.

___ **A** The larger cube has more mass than the smaller cube.

___ **B** The larger cube has less mass than the smaller cube.

___ **C** The larger cube melts at a higher temperature than the smaller cube.

___ **D** The larger cube melts at a lower temperature than the smaller cube.

___ **E** The density of the larger cube is greater than the smaller cube.

___ **F** The density of the larger cube is less than the smaller cube.

___ **G** The larger cube is more likely to float in water than the smaller cube.

___ **H** The larger cube is more likely to sink in water than the smaller cube.

___ **I** The larger cube is made up of larger atoms than the smaller cube.

___ **J** The larger cube is made up of smaller atoms than the smaller cube.

Explain your thinking. Describe the "rule" or reasoning you used to compare the cubes.

_____

_____

_____

_____

_____

Comparing Cubes

Teacher Notes

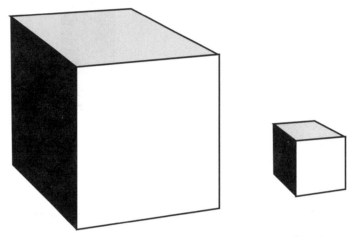

Purpose

The purpose of this assessment probe is to elicit students' ideas about intensive and extensive properties of matter. The probe is designed to find out which properties students think will change if the amount of material changes and which will stay the same.

Related Concepts

atoms or molecules, characteristic properties, density, extensive properties of matter, intensive properties of matter, mass, melting point, sinking and floating, weight

Explanation

The best response is A. The larger cube contains a greater amount of the same material, so its mass is greater. Mass is an extensive property that depends on the amount of material. Melting point and size of atoms are in-tensive properties that do not vary with the amount of material, so the melting point and size of atoms would remain the same. Density is also an intensive property. Expressed as a ratio of the mass to volume, the proportion of mass and volume remains constant when comparing cubes of different sizes that are made out of the same material. Since the de-gree to which a solid object floats in water de-pends on the density of the material and the two cubes have the same density, the larger cube is not more likely to float or sink in wa-ter than the smaller cube.

Curricular and Instructional Considerations

Elementary Students

At the elementary level, students describe observable properties of objects, including their size, weight, and ability to float or sink. Mass is a concept that is not introduced until later in elementary grades or in middle school. The idea that weight increases with the size of an object and that objects made of the same material float in a similar way is observable and is a grade-level expectation in the national standards. With modifications this probe may be useful in determining students' early, preconceived ideas about some of the properties of materials and ideas about atoms, but the concept of characteristic properties, such as density, exceeds expectations at this grade level. Students at this grade level should also not be expected to know that the size of atoms remains the same when the object increases in size. Several of the concepts in this probe are related to developing ideas related to constancy and change, a unifying theme that cuts across the elementary science curriculum. For example, the size of the cube may change, but it will still float the same way.

Middle School Students

In middle school, instructional experiences with the properties of matter progress from observational to conceptual. The national standards suggest that a distinction between extensive properties such as size, mass, or weight and intensive properties such as density and melting point can be made at this level. The term *characteristic property* is introduced at this level and is a grade-level expectation in the national standards. Students learn that characteristic properties are useful in identifying and comparing different substances. They begin

to distinguish characteristic properties from other observable properties they investigated in the elementary grades.

Students may use technical vocabulary such as *density* and *melting point* and may be able to calculate or measure a physical property, but it is important to determine if they have a conceptual understanding that these properties remain the same regardless of the size of the sample. Density is a particularly difficult concept at this level. An understanding of density progresses from the float and sink observations in the elementary grades to a proportional relationship between mass and volume at the middle school level. This probe is useful in determining if middle school students can make the link between floating and sinking and density. If their answers differ between responses E/F and G/H, it is important to probe further to understand why.

The notion of atoms is still abstract for many students at this level. The probe is useful in determining whether students have preconceived ideas about atoms and if they relate a macroscopic change in a substance (volume) to a microscopic change (increase in size of atoms).

High School Students

Instruction at the high school level builds on the concept of characteristic properties of substances that was developed in middle school and integrates the details of atomic structure with how atomic architecture plays a role in determining the properties of materials. This probe is useful in determining if students are able to explain the distinction between intensive and extensive properties at a substance or particle level. The probe may reveal that high school students revert to their preconceptions even after being taught the concept of characteristic properties in middle school.

Administering the Probe

Be sure students understand that the cubes are solid so that their notion that an object must have air in it to float does not interfere. It may help to have visual props for this probe, such as two different sizes of wood blocks. Make sure that students do not focus on the particular type of material but understand that the probe applies to any type of material, as long as both cubes are made of the same material (have the same composition).

Elementary teachers may wish to substitute the word *weight* for *mass* and remove distracters E/F and I/J if the terms and concepts are unfamiliar to students. Middle and high school teachers may want to add an additional set of distracters for other characteristic properties such as solubility and the extrinsic property of volume.

This probe may be combined with "Floating Logs" (p. 27), "Solids and Holes" (p. 41), "Turning the Dial" (p. 47), "Boiling Time and Temperature" (p. 53), and "Freezing Ice" (p. 59) to probe further for students' ideas related to characteristic properties.

Related Ideas in *National Science Education Standards* (NRC 1996)

. . . .

K–4 Properties of Objects and Materials

- Objects have many observable properties, including size, weight, shape, color, temperature, and the ability to react with other substances. Those properties can be measured using tools such as rulers, balances, and thermometers.

5–8 Properties and Changes in Properties of Matter

★ A substance has characteristic properties, such as density, a boiling point, and solubility, all of which are independent of the amount of sample. A mixture of substances often can be separated into the original substances using one or more of the characteristic properties.

Related Ideas in *Benchmarks for Science Literacy* (AAAS 1993)

. . . .

K–2 Structure of Matter

- Objects can be described in terms of the materials they are made of (clay, cloth, paper, etc.) and their physical properties (color, size, shape, weight, texture, flexibility, etc.).

K–2 Constancy and Change

★ Things change in some ways and stay the same in some ways.
- Things can change in different ways, such as size, weight, color, and movement.

3–5 Constancy and Change

★ Indicates a strong match between the ideas elicited by the probe and a national standard's learning goal.

★ Some features of things may stay the same even when other features change.

6–8 Structure of Matter

- Equal volumes of different substances usually have different weights.

9–12 Constancy and Change

- Things can change in detail but remain the same in general.

Related Research

- A study by Smith, Carey, and Wiser (1984) found that students' earliest ideas about density may be described by the phrase "heavy for its size." However, they fail to bring together the two ideas of size and "felt weight" so that density and weight are not differentiated but rather are included in a general notion of "heaviness" (Driver et al. 1994).

- Ideas that interfere with students' conception of density include the belief that when you change the shape of something you change its mass and the belief that heaviness is the most important factor in determining whether an object will sink or float (Stepans 2003).

- Although some students ages 14–22 relate density to compactness of particles, incomplete explanations may be due to their conceptions of mass and volume, which require understanding of the arrangement, concentration, and mass of particles (Driver et al. 1994).

- Many students have misconceptions about volume that present difficulties for understanding density (Driver et al. 1994).

- Due to an intuitive rule of "more A, more B," some students reason that if you have more material, properties such as melting point or density increase (Stavy and Tirosh 2000).

- Many students age 15 and over still use sensory reasoning about matter, despite being well advanced in thinking logically in other areas, such as mathematics (Barker 2004).

- Students generally do not regard a change in state, such as solid to liquid (melting), as being related to a specific temperature (Driver et al. 1994).

- Students of all ages show a wide range of ideas about particles. Many students will attribute macroscopic properties to particles (AAAS 1993). For example, they may believe that the size of the atoms that make up the cube increases as the size of the cube increases.

Suggestions for Instruction and Assessment

- This probe can be followed up with an investigation. Predict, observe, measure, and discuss what is the same and what is different about the mass (or weight), melting point, density, and flotation of two different-sized cubes of a substance (such as ice) that has a melting point that can be measured safely by middle and high school students.

- Provide multiple and varied opportunities for middle school students to observe and measure characteristic properties such as boiling point, melting point, density, and solubility using different amounts of the same substance.

- Have students test the middle school idea that "equal volumes of different substances usually have different weights" (AAAS 1993, p. 78). Have them test the corollary that equal weights (or masses if students are using a mass balance) of different substances usually have different volumes. Help students relate each to a conceptual understanding of density, constructing their own understanding of the $D = M/V$ relationship (density equals mass divided by volume) before being given the mathematical equation.

- Be aware that teaching a specific characteristic property such as density by itself may not help students develop a unified idea of characteristic properties that includes density, boiling point, melting point, and solubility. Be explicit in developing and pointing out the idea that all these properties have something in common—they do not depend on the amount of the sample.

- Conduct an investigation to determine the melting point of a small, medium, and large amount of the same substance, such as ice or wax (using appropriate safety precautions).

- Change the context to liquids—a small amount of liquid and a large amount of liquid. Ask similar questions, changing melting point to boiling point, and remove the distracters on floating (G/H). Have students safely test their ideas with substances such as water.

- To develop the distinction between characteristic and noncharacteristic properties (intensive and extensive), hold a mystery object in your hand (closed). Ask students if they can tell you what the object is if you give them the weight or mass, color, shape, texture, length, or other noncharacteristic properties that students readily observe in elementary grades. Elicit ideas about what kinds of properties might be helpful to know in order to identify the mystery object. After developing the idea of characteristic properties through a variety of instructional experiences, revisit the mystery object in your closed hand. Ask the same questions about which properties would help them identify what the mystery object is. Use the information formatively to assess whether students have grasped an understanding of characteristic properties.

- Have students practice using "if, then" reasoning in the context of physical properties. For example, prompt them to respond to statements such as, "If the volume of a substance increases, then its boiling point ____ because ____." This can be practiced with elementary students using basic properties of objects; for example, "If the shape of the clay ball changes, its weight will ____ because ____."

Related NSTA Science Store Publications and NSTA Journal Articles

American Association for the Advancement of Science (AAAS). 1993. *Benchmarks for science literacy.* New York: Oxford University Press.

Driver, R., A. Squires, P. Rushworth, and V. Wood-Robinson. 1994. *Making sense of secondary science: Research into children's ideas.* London: RoutledgeFalmer.

Keeley, P. 2005. *Science curriculum topic study: Bridging the gap between standards and practice.* Thousand Oaks, CA: Corwin Press.

Libarkin, J., C. Crockett, and P. Sadler. 2003. Density on dry land: Demonstrations without buoyancy challenge student misconceptions. *The Science Teacher* 70 (6): 46–50.

National Research Council (NRC). 1996. *National science education standards.* Washington, DC: National Academy Press.

Peterson-Chin, L., and D. Sterling. 2004. Looking at density from different perspectives. *Science Scope* 27 (7): 16–20.

Shaw, M. 1998. Diving into density. *Science Scope* 22 (3): 24–26.

Stepans, J. 2003. *Targeting students' science misconceptions: Physical science concepts using the conceptual change model.* Tampa, FL: Idea Factory.

Talanquer, V. 2002. Minimizing misconceptions: Tools for identifying patterns of reasoning. *The Science Teacher* 69 (8): 46–49.

References

American Association for the Advancement of Science (AAAS). 1993. *Benchmarks for science literacy.* New York: Oxford University Press.

Barker, V. 2004. *Beyond appearances: Students' misconceptions about basic chemical ideas. A report prepared for the Royal Society of Chemistry.* Cambridge, England.

Driver, R., A. Squires, P. Rushworth, and V. Wood-Robinson. 1994. *Making sense of secondary science: Research into children's ideas.* London: RoutledgeFalmer.

Keeley, P. 2005. *Science curriculum topic study: Bridging the gap between standards and practice.* Thousand Oaks, CA: Corwin Press.

National Research Council (NRC). 1996. *National science education standards.* Washington, DC: National Academy Press.

Smith, C., S. Carey, and M. Wiser. 1984. A case study of the development of size, weight, and density. *Cognition* 21 (3): 177–237.

Stavy, R., and D. Tirosh. 2000. *How students (mis-) understand science and mathematics: Intuitive rules.* New York: Teachers College Press.

Stepans, J. 2003. *Targeting students' science misconceptions: Physical science concepts using the conceptual change model.* Tampa, FL: Idea Factory.

Related Curriculum Topic Study Guides

(Keeley 2005)

"Density"

"Physical Properties and Change"

"Properties of Matter"

Floating Logs

A log was cut from a tree and put in water. The log floated on its side so that half the log was above the water surface. Another log was cut from the same tree. This log was twice as long and twice as wide. How does the larger log float compared with the smaller log? Circle the best answer:

A More than half of the larger log floats above the water surface.

B Half of the larger log floats above the water surface.

C Less than half of the larger log floats above the water surface.

Explain your thinking. Describe the "rule" or the reasoning you used for your answer.

_____

_____

_____

_____

_____

_____

_____

_____

_____

_____

_____

Floating Logs

Teacher Notes

Purpose

The purpose of this assessment probe is to elicit students' ideas about density. The probe is designed to find out if students think changing the size of an object affects its density.

Related Concepts

characteristic properties, density, intensive properties of matter, sinking and floating.

Explanation

The best response is B: Half of the larger log floats above the water surface. The degree to which a solid object will float when placed in water depends on the density of the material. When a second object is compared with a floating solid, a solid object with a lesser density will float higher above the water's surface, an object with the same density will float at

equal levels, and an object with a greater density will be more submerged. Density is a characteristic property of matter, which means that it is independent of the amount of material. If one sample of material is very large and another sample of the same material is very small, the proportion (ratio) of the mass to volume of each sample is still the same, so the density remains the same. The first and second logs were both cut from the same tree, so they are made of the same material and have close to the same density. (There may be a slight difference because the logs are not made of a homogeneous material.) Since the densities are for practical purposes the same, the two different-sized logs will float at equal levels. One-half of the first log floated above the water's surface, so one-half of the second (larger) log will also float above the water's surface.

Curricular and Instructional Considerations

Elementary Students

At the elementary level, students have observational experiences with floating and sinking objects of different sizes and shapes. They are able to describe observable properties of objects, such as how much of an object floats above the water's surface. They begin to develop an understanding of the unifying theme of constancy and change: Even though some things may change (such as size), other things may stay the same (ability to float). This probe may be useful in determining students' ideas about floating objects, but the concept of characteristic properties, such as density, should wait until middle school.

Middle School Students

In middle school, instructional experiences with density progress from observational (floating or sinking and heavy for its size) to a conceptual understanding of density as a characteristic property of matter. Students begin to use mathematics to quantitatively describe density. The national standards suggest that middle school is the time to make the distinction between extrinsic properties such as size, mass, or weight and characteristic properties such as density. By the end of middle school students should understand that two objects composed of the same substance will have the same characteristic properties, which can be used to identify them or

predict their behavior. Students begin to use technical vocabulary such as *mass, volume,* and *density.* However, it is important to determine if they have a conceptual understanding of density before introducing the *D = M/V* relationship (density equals mass divided by volume).

High School Students

Density experiences at the high school level include symbolic representations of density using the variables of mass and volume to calculate proportional relationships. Applications of density are extended to an understanding of the Earth, astronomy, life science, and the designed world. However, a conceptual understanding of density still eludes many high school students.

Administering the Probe

You may wish to use props to help younger students visualize the manner in which the first log is floating with respect to the water's surface and to show students what it means when logs float on their sides, rather than upright like a buoy. Place an object that floats in a clear container of water so that students can see what is meant by "above and below the water's surface" and "floating on its side," or draw a picture to explain it. Show students a second object composed of the same material that is longer and wider than the first object, but don't place this object in the water.

The probes "Comparing Cubes" (p. 19) and "Solids and Holes" (p. 41) can also be used to determine if students recognize that density is a characteristic property of matter.

Related Ideas in *National Science Education Standards* (NRC 1996)

. . . .

K–4 Properties of Objects and Materials

- Objects have many observable properties, including size, weight, shape, color, temperature, and the ability to react with other substances. Those properties can be measured using tools such as rulers, balances, and thermometers.

5–8 Properties and Changes in Properties of Matter

- ★ A substance has characteristic properties, such as density, a boiling point, and solubility, all of which are independent of the amount of sample. A mixture of substances often can be separated into the original substances using one or more of the characteristic properties.

Related Ideas in *Benchmarks for Science Literacy* (AAAS 1993)

K–2 Structure of Matter

- Objects can be described in terms of the materials they are made of (e.g., clay, cloth, paper) and their physical properties (e.g., color, size, shape, weight, texture, flexibility).

K–2 Constancy and Change

- ★ Things change in some ways and stay the same in some ways.

3–5 Constancy and Change

- ★ Some features of things may stay the same even when other features change.

6–8 Structure of Matter

- Equal volumes of different substances usually have different weights.

9–12 Constancy and Change

- Things can change in detail but remain the same in general.

Related Research

- Ideas that interfere with students' conception of density include the belief that when you change the shape of something you change its mass and the belief that heaviness is the most important factor in determining whether an object will sink or float (Stepans 2003).
- Many students have misconceptions about volume that present difficulties for understanding density (Driver et al. 1994).
- Driver et al. (1994) described a study conducted by Biddulph and Osborne (1984) in which some students ages 7–14 suggested that things float because they are light and, when asked why objects float, offered different reasons for different objects. The same study asked children ages 8–12 how a longer candle would float compared with a shorter piece; many students thought the longer candle would sink/float lower (Driver et al. 1994).

★ Indicates a strong match between the ideas elicited by the probe and a national standard's learning goal.

- Some students use an intuitive rule of "more A, more B." They reason that if you have more material, density increases or makes an object sink more (Stavy and Tirosh 2000).
- Students' ways of looking at floating and sinking include the roles played by material, weight, shape, cavities, holes, air, and water (Driver et al. 1994).

Suggestions for Instruction and Assessment

- This probe can be followed up with an inquiry-based investigation using wooden dowels of different lengths and thicknesses.
- Try a very small piece of Ivory soap (a soap that floats) versus the rest of the bar of Ivory soap. Or use a soap that sinks, cut off a tiny piece, and ask students if they think that piece of soap will float, sink, or float differently depending on its size. Sometimes students think that a tiny piece of an object will behave differently from the whole object.
- Investigate the floating and sinking of the same kind of material—for example, Styrofoam balls—with different sizes and the same shape. Similar investigations can be conducted with strawberries, blocks of wood, or rubber objects.
- Investigate the floating and sinking of the same kind of material made of different shapes. For example, would a block of Styrofoam float the same way as a Styrofoam sphere? Again, this investigation can be conducted with other materials.

- When middle school or high school students have developed the conceptual understanding of density, have them use mathematics to support their explanations with proportional reasoning. It is counterproductive to start by using $D = M/V$ if students have not developed a conceptual understanding of density first.

Related NSTA Science Store Publications and NSTA Journal Articles

American Association for the Advancement of Science (AAAS). 1993. *Benchmarks for science literacy.* New York: Oxford University Press.

Driver, R., A. Squires, P. Rushworth, and V. Wood-Robinson. 1994. *Making sense of secondary science: Research into children's ideas.* London: RoutledgeFalmer.

Keeley, P. 2005. *Science curriculum topic study: Bridging the gap between standards and practice.* Thousand Oaks, CA: Corwin Press.

Libarkin, J., C. Crockett, and P. Sadler. 2003. Density on dry land: Demonstrations without buoyancy challenge student misconceptions. *The Science Teacher* 70 (6): 46–50.

National Research Council (NRC). 1996. *National science education standards.* Washington, DC: National Academy Press.

Peters, E. 2005. Reforming cookbook labs. *Science Scope* 29 (3): 16–21.

Peterson-Chin, L., and D. Sterling. 2004. Looking at density from different perspectives. *Science Scope* 27 (7): 16–20.

Sampson, V. 2006. Two-tiered assessment. *Science Scope* 29 (5): 46–49.

Shaw, M. 1998. Diving into density. *Science Scope* 22 (3): 24–26.

Stepans, J. 2003. *Targeting students' science misconceptions: Physical science concepts using the conceptual change model.* Tampa, FL: Idea Factory.

Vanides, J., Y. Yin, M. Tomita, and M. Ruiz-Primo. 2005. Using concept maps in the science classroom. *Science Scope* 28 (8): 27–31.

Related Curriculum Topic Study Guides
(Keeley 2005)
"Density"
"Physical Properties and Change"

References

American Association for the Advancement of Science (AAAS). 1993. *Benchmarks for science literacy.* New York: Oxford University Press.

Biddulph, F., and R. Osborne. 1984. Pupils' ideas about floating and sinking. Paper presented to the Australian Science Education Research Association Conference, Melbourne, Australia.

Driver, R., A. Squires, P. Rushworth, and V. Wood-Robinson. 1994. *Making sense of secondary science: Research into children's ideas.* London: RoutledgeFalmer.

Keeley, P. 2005. *Science curriculum topic study: Bridging the gap between standards and practice.* Thousand Oaks, CA: Corwin Press.

National Research Council (NRC). 1996. *National science education standards.* Washington, DC: National Academy Press.

Stavy, R., and D. Tirosh. 2000. *How students (mis-) understand science and mathematics: Intuitive rules.* New York: Teachers College Press.

Stepans, J. 2003. *Targeting students' science misconceptions: Physical science concepts using the conceptual change model.* Tampa, FL: Idea Factory.

Floating High and Low

Sam put a solid ball in a tank of water. As shown by the ball on the left, it floated halfway above and halfway below the water level. What can Sam do to make a ball float like the ball on the right? Put an X next to all the things Sam can do to have a solid ball float so that most of it is below the water level.

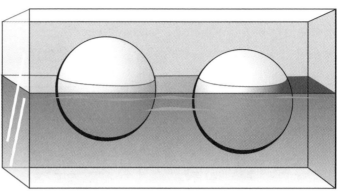

____ **A** Use a larger ball made out of the same material.

____ **B** Use a smaller ball made out of the same material.

____ **C** Use a ball of the same size made out of a denser material.

____ **D** Use a ball of the same size made out of less dense material.

____ **E** Add more water to the tank so it is deeper.

____ **F** Add salt to the water.

____ **G** Attach a weight to the ball.

Explain your thinking. Describe the "rule" or reasoning you used to determine how to change how an object floats in water.

_____

_____

_____

_____

_____

_____

_____

Floating High and Low

Teacher Notes

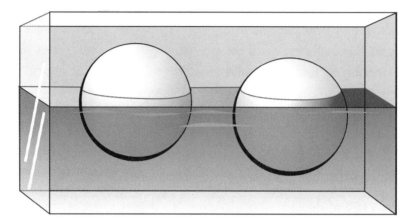

Purpose

The purpose of this assessment probe is to elicit students' ideas about density and buoyancy. The probe is designed to find out how students think an object can be made to float differently.

Related Concepts

buoyancy, characteristic properties, density, intensive properties of matter, sinking and floating

Explanation

The best responses are C and G. To make the solid ball float so that most of it is under the water, you can either use a ball of the same size made out of a denser material or attach a weight to the ball. The degree to which a solid object will float when placed in water depends on the density of the material. To be further submerged, the density of the object must be increased. Density is defined as the ratio of the mass to the volume of an object. By using a ball of the same size made out of a denser material, the ratio of the mass to volume is greater and the object will be further submerged. By taping a weight to the ball, the proportion of the total mass relative to volume is increased, and so the overall density is increased. This, too, will result in the object being further submerged. As more matter is attached to the ball, the buoyant force increases, indicated by the displacement of more water.

Adding more water to the tank makes no difference in how an object floats. An object floats the same way regardless of how deep or shallow the water is. Adding salt to the water actually makes the object more buoyant because salt increases the density of the water. For example, when you swim in the ocean, you float better than when you swim in freshwater because saltwater is denser than freshwater.

Curricular and Instructional Considerations

Elementary Students

At the elementary level, students typically have experiences with floating and sinking objects of different sizes and shapes. They are able to describe observable properties of objects. Although it is too early to expect them to quantitatively explain results, they can make changes to objects and observe how the change affects how they float. Upper elementary students may be more systematic in their investigation of floating and sinking objects because they most likely have had opportunities to measure weight or mass and volume and begin to develop experiments that involve variables as fair tests of their ideas. This probe may be useful in determining students' preconceptions about floating and sinking, but the concepts of characteristic properties, quantitative density, and buoyant forces are beyond this grade level.

Middle School Students

In middle school, observational experiences that involve floating and sinking progress to instructional opportunities at a conceptual level involving density. By the end of middle school, a mathematical approach toward reasoning about density is incorporated. The national standards suggest that a distinction between

extensive properties such as size or weight and characteristic or intensive properties such as density should be made at this grade level. Students may use technical vocabulary such as *mass, volume,* and *density,* but it is important to determine if they have a conceptual understanding of the proportional nature of density. Students also develop ideas about pairs of forces in fluids, such as the buoyant force that pushes an object upward and the gravitational force that pulls it downward.

This probe is useful in determining if students can explain the significance of the characteristic property of density in relation to changing how an object floats: if an object's mass relative to its volume is increased, its density will increase. It can also be used to see if students recognize the effect of opposing forces when weight is added to a floating object placed in a fluid.

High School Students

Instructional opportunities at the high school level include symbolic representations of density, using the variables of mass and volume to calculate proportional relationships. Mathematical applications of density are extended to the living and designed world. This probe is useful in determining if students are able to apply mathematical reasoning to a concrete application as well as use the concept of opposing forces affecting a buoyant object. This probe is also useful in determining whether students hold on to their preconceptions about density and floating objects years after instructional opportunities to learn about density and density-related phenomena.

Administering the Probe

It may help to have visual props for this probe. Place a sphere that floats in a container of water. Then display objects and materials that represent each of the possibilities as students respond and explain their thinking. Remove distracters that include terminology that is unfamiliar to younger students or simplify the terminology; for example, replace "made out of a denser material" with "made out of material that is heavier for its size."

Teachers may want to continue to probe students' ideas about density or properties of objects and materials by using this probe in conjunction with the probes "Comparing Cubes" (p. 19), "Floating Logs" (p. 27), or "Solids and Holes" (p. 41).

Related Ideas in *National Science Education Standards* (NRC 1996)

· ·

K–4 Properties of Objects and Materials

- Objects have many observable properties, including size, weight, shape, color, temperature, and the ability to react with other substances. Those properties can be measured using tools such as rulers, balances, and thermometers.

5–8 Properties and Changes in Properties of Matter

- ★ A substance has characteristic properties, such as density, a boiling point, and solu-

bility, all of which are independent of the amount of sample. A mixture of substances often can be separated into the original substances using one or more of the characteristic properties.

Related Ideas in *Benchmarks for Science Literacy* (AAAS 1993)

· ·

K–2 Structure of Matter

- Objects can be described in terms of the materials they are made of (clay, cloth, paper, etc.) and their physical properties (color, size, shape, weight, texture, flexibility, etc.).

K–2 Constancy and Change

- ★ Things change in some ways and stay the same in some ways.

3–5 Constancy and Change

- ★ Some features of things may stay the same even when other features change.

6–8 Structure of Matter

- Equal volumes of different substances usually have different weights.

6–8 Constancy and Change

- ★ Symbolic equations can be used to summarize how the quantity of something changes over time or in response to other changes.

★ Indicates a strong match between the ideas elicited by the probe and a national standard's learning goal.

9–12 Constancy and Change

- Things can change in detail but remain the same in general.

Related Research

- Students' earliest ideas about density may be expressed in terms of "heavy for its size" (Driver et al. 1994).

- Children compare objects by their "felt weight" and over time develop the idea that felt weight is characteristic of certain objects or materials. The idea of mass and the use of the word are a concept and terminology that need to develop over time. Younger students often associate mass with the similar-sounding familiar word *massive* and thus confuse mass with size or volume. Hence students estimate mass of an object or material from its bulk appearance (Driver et al. 1994).

- Ideas that interfere with students' conception of density include the belief that when you change the shape of something you change its mass and the belief that heaviness is the most important factor in determining whether an object will sink or float (Stepans 2003).

- Notions of weight and density develop as children begin to take account of viewpoints other than their own. At ages 9–10, students begin to relate density of one material to another. For example, they may say a material floats because it is "lighter than water" (Driver et al. 1994).

- Although some students ages 14–22 relate density to compactness of particles, in-complete explanations may be due to their conceptions of mass and volume, which require understanding of the arrangement, concentration, and mass of particles (Driver et al. 1994).

- Many students have misconceptions about volume that present difficulties for understanding density (Driver et al. 1994).

- Driver et al. (1994) described a study conducted by Biddulph and Osborne (1984) in which some students ages 7–14 suggested that things float because they are light and, when asked why objects float, offered different reasons for different objects. The same study asked children ages 8–12 how a longer candle would float compared with a shorter piece; many students thought the longer candle would sink/float lower. By ages 9–10, many children recognized that the depth of water would not affect how an object floats; however, even at age 11 and up, there were still children who thought otherwise (Driver et al. 1994).

- Some students use an intuitive rule of "more A, more B" to reason that if you have more material, density increases (Stavy and Tirosh 2000).

- Often students will mistake buoyancy-related phenomena for characteristics of density (Libarkin, Crockett, and Sadler 2003).

- When students investigate and explain sinking and floating, they typically focus only on the object they are testing and ignore the liquid that the object is in (Houghton et al. 2000).

Suggestions for Instruction and Assessment

- This probe can be followed up with an inquiry-based investigation. Have students predict, observe, systematically test, and explain how solid objects will float in water when changes are made to them or the liquid.

- This probe lends itself nicely to a station approach to investigate floating and sinking phenomena. Have students make a prediction about things that can be done to change how an object floats based on the choices given in the probe. Students can then test their ideas using a variety of stations set up around the room, including wooden or Styrofoam balls of different sizes; same-sized balls made of different materials; a container in which students can change the depth of water a ball is put into; a ball with the water level marked to show where it floats in tap water (ask students to add increasing amounts of salt to the container of water and observe what happens to the mark where the water level was); a variety of weights that can be attached to a floating ball (ask students to try to get the ball to float like the second picture on the probe).

- When dealing with density-related phenomena, use the terminology *mass, volume,* and *density* with older students. With young children, use the more familiar terms *size, weight,* and *heavy for its size* instead of *density*. Research indicates that young students mistake the word *mass* for *massive* and confuse it with the size of objects. In other words, a large Styrofoam ball is more "massive" to them than a small wooden ball.

- Change the object. Have students investigate whether cutting a banana into a variety of shapes and sizes will change how it floats in water. Put an orange in water and then compare how it floats after the peel is removed.

Related NSTA Science Store Publications and NSTA Journal Articles

American Association for the Advancement of Science (AAAS). 1993. *Benchmarks for science literacy.* New York: Oxford University Press.

Driver, R., A. Squires, P. Rushworth, and V. Wood-Robinson. 1994. *Making sense of secondary science: Research into children's ideas.* London: RoutledgeFalmer.

Keeley, P. 2005. *Science curriculum topic study: Bridging the gap between standards and practice.* Thousand Oaks, CA: Corwin Press.

Libarkin, J., C. Crockett, and P. Sadler. 2003. Density on dry land: Demonstrations without buoyancy challenge student misconceptions. *The Science Teacher* 70 (6): 46–50.

National Research Council (NRC). 1996. *National science education standards.* Washington, DC: National Academy Press.

Robertson, W. 2005. *Air, water, and weather: Stop faking it! Finally understanding science so you can teach it.* Arlington, VA: NSTA Press.

Shaw, M. 1998. Diving into density. *Science Scope* 22 (3): 24–26.

Stepans, J. 2003. *Targeting students' science misconceptions: Physical science concepts using the conceptual change model.* Tampa, FL: Idea Factory.

Related Curriculum Topic Study Guide

(Keeley 2005)

"Density"

References

American Association for the Advancement of Science (AAAS). 1993. *Benchmarks for science literacy.* New York: Oxford University Press.

Biddulph, F., and R. Osborne. 1984. Pupils' ideas about floating and sinking. Paper presented to the Australian Science Education Research Association Conference, Melbourne, Australia.

Driver, R., A. Squires, P. Rushworth, and V. Wood-Robinson. 1994. *Making sense of secondary science: Research into children's ideas.* London: RoutledgeFalmer.

Houghton, C., K. Record, B. Bell, and T. Grotzer. 2000. Conceptualizing density with a relational systemic model. Paper presented at the annual conference of the National Association for Research in Science Teaching, New Orleans, LA.

Keeley, P. 2005. *Science curriculum topic study: Bridging the gap between standards and practice.* Thousand Oaks, CA: Corwin Press.

Libarkin, J., C. Crockett, and P. Sadler. 2003. Density on dry land: Demonstrations without buoyancy challenge student misconceptions. *The Science Teacher* 70 (6): 46–50.

National Research Council (NRC). 1996. *National science education standards.* Washington, DC: National Academy Press.

Stavy, R., and D. Tirosh. 2000. *How students (mis-) understand science and mathematics: Intuitive rules.* New York: Teachers College Press.

Stepans, J. 2003. *Targeting students' science misconceptions: Physical science concepts using the conceptual change model.* Tampa, FL: Idea Factory.

Solids and Holes

Lance had a thin, solid piece of material. He placed the material in water and it floated. He took the material out and punched holes all the way through it. What do you think Lance will observe when he puts the material with holes back in the water? Circle your prediction.

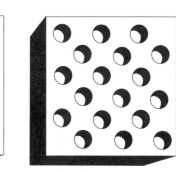

A It will sink.

B It will barely float.

C It will float the same as it did before the holes were punched in it.

D It will neither sink nor float. It will bob up and down in the water.

Explain your thinking. Describe the "rule" or reasoning you used to make your prediction.

_____

_____

_____

_____

_____

_____

_____

_____

_____

_____

_____

Solids and Holes

Teacher Notes

 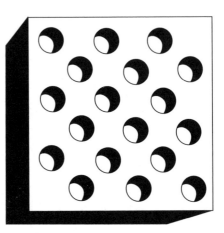

Purpose

The purpose of this assessment probe is to elicit students' ideas about density. The task is designed to find out if students think a solid object will float differently if it has holes poked all the way through it or if they confuse it with mixed density.

Related Concepts

Characteristic properties, density, intensive properties of matter, sinking and floating

Explanation

The best response is C: It will float the same as it did before the holes were punched in it. The degree to which a solid object will float when placed in water depends on the density of the material. *Density,* a characteristic property of

matter, is defined as the ratio of the mass to the volume of an object. This intensive property is independent of the amount of material. As holes are punched uniformly throughout the object, regardless of what the material is, the amount of mass and volume that is removed is proportional, so the remaining amount of material will have the same density. Since the density remains the same, the object will continue to float in the same manner.

In contrast, *mixed density* is a condition in which there is more than one substance making up the volume of an object. For example, a hollow plastic ball may contain air in it, which would give it a mixed density that includes the density of the plastic and the density of air. The example provided in this probe is not one of mixed density because the object is solid and the holes that are drilled go all the way through and are not trapping air inside the object, such as in a steel-hulled boat.

Curricular and Instructional Considerations

Elementary Students

At the elementary level, students typically have experiences with floating and sinking objects of different sizes and shapes and are able to describe observable properties of objects. They do things to change the shape of objects to make them float. They also have experiences in floating objects that contain air as well as everyday knowledge about floating objects such as boats. Because of their experiences with mixed density (objects that contain air in addition to solid material), such as floating boats, they may not recognize that it is the displacement of air by water that causes a boat or object to sink rather than the fact that it has a hole in it. This probe may be useful in determining students' preconceptions about floating and sinking when holes are punched through floating material, but the concepts of characteristic properties and quantitative density are beyond this grade level.

Middle School Students

In middle school, observational experiences of floating and sinking progress to instructional opportunities at a conceptual level involving density. Students develop an understanding of density of a substance and how

that differs from mixed density. By the end of middle school, a mathematical approach toward reasoning about density is incorporated, including proportional reasoning. The national standards suggest that a distinction between extensive properties such as size or weight and intensive or characteristic properties such as density should be made at this grade level.

Students may use technical vocabulary such as *mass, volume,* and *density,* but it is important to determine if they have a conceptual understanding of the proportional nature of density. This probe is useful in determining if students can explain the significance of the characteristic property of density in relation to changing how an object floats: If an object's mass relative to its volume is unchanged, then its density remains constant.

High School Students

Instructional opportunities at the high school level include symbolic representations of density, using the variables of mass and volume to calculate proportional relationships. Mathematical applications of density are extended to the living and designed world and include examples of mixed density in which students can explain the difference in density caused by the arrangement of air molecules contained within a "solid" substance. This probe is useful in determining if students are able to apply mathematical reasoning in a concrete context.

Administering the Probe

It may help to have visual props for this probe. Place a Styrofoam block or thin piece of wood in a container of water to show students how it floats. Then use a coring device to drill holes that go all the way through the block, but don't place it in the water. Ask students to commit to a prediction once they see how the object has changed. Teachers may want to continue to probe for density-related ideas using the "Comparing Cubes" (p. 19), "Floating Logs" (p. 27), and "Floating High and Low" (p. 33) probes.

Related Ideas in *National Science Education Standards* (NRC 1996)

. . . .

K–4 Properties of Objects and Materials

- Objects have many observable properties, including size, weight, shape, color, temperature, and the ability to react with other substances. Those properties can be measured using tools such as rulers, balances, and thermometers.

5–8 Properties and Changes in Properties of Matter

- ★ A substance has characteristic properties, such as density, a boiling point, and solubility, all of which are independent of the amount of the sample. A mixture of substances often can be separated into the original substances using one or more of the characteristic properties.

Related Ideas in *Benchmarks for Science Literacy* (AAAS 1993)

. . . .

K–2 Structure of Matter

- Objects can be described in terms of the materials they are made of (e.g., clay, cloth, paper) and their physical properties (e.g., color, size, shape, weight, texture, flexibility).

K–2 Constancy and Change

- ★ Things change in some ways and stay the same in some ways.

3–5 Constancy and Change

- ★ Some features of things may stay the same even when other features change.

6–8 Structure of Matter

- Equal volumes of different substances usually have different weights.

9–12 Constancy and Change

- Things can change in detail but remain the same in general.

Related Research

- Some students think that for an object to float, it must contain air (Stepans 2003).
- Ideas that interfere with students' conception of density include the belief that when you change the shape of something, you change its mass, and that heaviness is the most important factor in determining whether an object will sink or float (Stepans 2003).

★ Indicates a strong match between the ideas elicited by the probe and a national standard's learning goal.

- Many students have misconceptions about volume that present difficulties for understanding density (Driver et al. 1994).

- Some students may use an intuitive rule of "more A, more B" or its converse "less A, less B" to reason that if you have less material, the ability to float decreases (Stavy and Tirosh 2000).

- Often students will mistake buoyancy-related phenomena for characteristics of density (Libarkin, Crockett, and Sadler 2003).

- In a study by Grimillini, Gandolfi, and Pecori Balandi (1990) of children's ideas related to buoyancy, they found that children take into account four factors when considering how objects float: (1) the role played by material and weight; (2) the role played by shape, cavities, and holes; (3) the role played by air; and (4) the role played by water. Many children thought that holes in objects affected their ability to float. Even when taught that holes through solid objects do not change the object's ability to float, the idea was still firmly held by students (Driver et al. 1994).

Suggestions for Instruction and Assessment

- This probe can be followed up with an inquiry-based investigation. Ask students to predict, observe, test, and explain what happens when a thin piece of wood, Styrofoam, or other floating material has holes uniformly punched all the way through it. Students should test their ideas with a variety of materials to ensure that their ideas apply regardless of the type of material used to test their ideas.

- Test the same idea with an object. For example, test how an apple will behave in water when a hole is cored through the apple.

- Help students distinguish between mixed density and density of a pure substance. Have them reason why materials like metals can be made to float, comparing and contrasting a mixed-density object with a pure-density object. For example, a hollow metal sphere filled with air would float because a large part of the volume of the object is made up of air, which decreases its density, whereas a sphere of solid metal would sink because its density is greater than water.

- Once students have grasped the idea targeted by the probe, present them with a new situation. Contrast the "holes all the way through" model with the "cavity" model. Place an object such as soap (not Ivory) that sinks in water. Take the soap and carve or drill holes uniformly (not all the way through, but rather holes that form cavities) so that much of its volume will be displaced by air, causing it to float. Ask students to predict what will happen if you take the soap and continue making the holes so they go all the way through. Listen carefully to their ideas, noting if they base their prediction on the probe example, without considering that the object did not float to begin with. Have them test their prediction and consider an explanation for this counter example.

• At the middle and high school level, be sure to include density investigations that do not involve only objects floating and sinking in water. Such activities should confirm that the density of an object is independent of size, shape, and amount regardless of whether it is in water or in dry conditions.

Related NSTA Science Store Publications and NSTA Journal Articles

American Association for the Advancement of Science (AAAS). 1993. *Benchmarks for science literacy.* New York: Oxford University Press.

Driver, R., A. Squires, P. Rushworth, and V. Wood-Robinson. 1994. *Making sense of secondary science: Research into children's ideas.* London: RoutledgeFalmer.

Keeley, P. 2005. *Science curriculum topic study: Bridging the gap between standards and practice.* Thousand Oaks, CA: Corwin Press.

Libarkin, J., C. Crockett, and P. Sadler. 2003. Density on dry land: Demonstrations without buoyancy challenge student misconceptions. *The Science Teacher* 70 (6): 46–50.

National Research Council (NRC). 1996. *National science education standards.* Washington, DC: National Academy Press.

Shaw, M. 1998. Diving into density. *Science Scope* 22 (3): 24–26.

Stepans, J. 2003. *Targeting students' science misconceptions: Physical science concepts using the conceptual change model.* Tampa, FL: Idea Factory.

Related Curriculum Topic Study Guide
(Keeley 2005)
"Density"

References

American Association for the Advancement of Science (AAAS). 1993. *Benchmarks for science literacy.* New York: Oxford University Press.

Driver, R., A. Squires, P. Rushworth, and V. Wood-Robinson. 1994. *Making sense of secondary science: Research into children's ideas.* London: RoutledgeFalmer.

Grimillini, T., E. Gandolfi, and B. Pecori Balandi. 1990. Teaching strategies and conceptual change: Sinking and floating at elementary school level. Paper presented at the Australian Science Education Research Association Conference, Melbourne, Australia.

Keeley, P. 2005. *Science curriculum topic study: Bridging the gap between standards and practice.* Thousand Oaks, CA: Corwin Press.

Libarkin, J., C. Crockett, and P. Sadler. 2003. Density on dry land: Demonstrations without buoyancy challenge student misconceptions. *The Science Teacher* 70 (6): 46–50.

National Research Council (NRC). 1996. *National science education standards.* Washington, DC: National Academy Press.

Stavy, R., and D. Tirosh. 2000. *How students (mis-) understand science and mathematics: Intuitive rules.* New York: Teachers College Press.

Stepans, J. 2003. *Targeting students' science misconceptions: Physical science concepts using the conceptual change model.* Tampa, FL: Idea Factory.

Turning the Dial

Flora is boiling water on a stove. She turns the temperature dial up to high to boil the water. The water is boiling vigorously with large bubbles quickly forming and bursting at the surface. Flora then turns the dial of the stove down to low. The water is boiling gently, with smaller bubbles slowly forming and bursting at the surface. Flora wonders if the boiling temperature changes when she turns the dial. What would you tell Flora? Circle the best answer.

A The boiling temperature is greater when the dial is set at high.

B The boiling temperature is greater when the dial is set at low.

C The boiling temperature is the same at both settings.

Explain your thinking. What "rule" or reasoning did you use to select your answer?

_____

_____

_____

_____

_____

_____

_____

_____

_____

Turning the Dial

Teacher Notes

Purpose

The purpose of this assessment probe is to elicit students' ideas about boiling point. The probe is designed to find out whether students recognize that if a liquid is boiling under standard conditions, the boiling point remains constant, no matter how much heat is applied.

Related Concepts

boiling and boiling point, characteristic properties, change in state, energy, heat, intensive properties of matter, temperature

Explanation

The best response is C: The boiling temperature is the same at both settings. Temperature is a measure of the average kinetic energy of the molecules in a system. When heat is applied to a liquid, the kinetic energy of the molecules increases. The motion, and hence temperature, increases until the temperature of the liquid reaches its boiling point. (Each pure substance has a specific temperature at which it will boil.) Once a liquid is at its boiling point, the heat supplied to the system is used to overcome attractive forces between particles in the liquid. This results in conversion of molecules from a liquid to a gas phase, allowing the molecules to escape into the air. Although the amount of heat going into the system relative to the amount of energy flowing into the water as it changes from the liquid to gas in any given interval impacts the rate of change, as long as the liquid is boiling, the boiling temperature of the remaining liquid remains essentially the same.

Curricular and Instructional Considerations

. . . .

Elementary Students

At the elementary level, students' experiences with materials are primarily observational. The idea of change is connected to physical properties by subjecting materials to temperature changes through heating and freezing. Upper elementary students are familiar with the change in states of water from the solid to liquid to gas phase and vice versa. They learn how to use thermometers to measure the temperature of water. They may observe that water boils at 100°C, but the concept of the difference between heat and temperature is too complex for this age. However, the probe is useful in determining whether students intuitively believe that increasing the heat (e.g., turning up the dial) makes boiling water "hotter." They may have developed this idea early on through their everyday experience in the kitchen and hold on to this idea through school, even after being taught the concept.

Middle School Students

In middle school, students shift their focus from properties of materials to the characteristic properties of the substances from which the materials are made. Students learn about the characteristics of different states of matter and the properties associated with phase changes from solid to liquid to gas. Opportunities to observe and measure characteristic properties such as boiling and melting points can be used to distinguish and separate one substance from

another. Identifying characteristic properties, such as boiling point, is a grade-level expectation in the national standards. Students develop the idea that substances have a constant boiling point and that this boiling point does not change under standard conditions, no matter how vigorously or gently the liquid boils. Students are beginning to connect the ideas of heat, temperature, and constant boiling point, although the idea of a constant boiling temperature is counterintuitive, given their everyday experiences with boiling liquids in the kitchen.

High School Students

During high school, instructional opportunities connect the macroscopic properties of substances studied in middle school to microscopic properties. This probe may be useful in determining if students can relate the particulate nature of liquids and gases to the role of thermal energy in phase changes. It is a grade-level expectation that students can distinguish between heat and temperature and explain why boiling water stays at the same temperature regardless of how much heat is applied. However, the distinction between heat and temperature is still a concept that eludes many high school students.

Administering the Probe

It may be helpful to have visual props for this probe. Make sure students are familiar with the dial used to adjust burner temperature on a stove, since children may have experienced a variety of stoves and dials in their home settings. Bring a beaker of water to a vigorous boil. Lower the amount of heat supplied to the boiling water, and keep the water at a slow boil as students respond to the probe and explain their thinking.

Teachers may want to continue to probe reasoning about heat, energy, temperature, and states of matter by using this probe in conjunction with the probes "Boiling Time and Temperature" (p. 53) or "What's in the Bubbles?" (p. 65).

Suggestions for Teachers on Terminology

Heat and temperature are not the same thing. Heat is a measure of energy flow, whereas temperature is a macroscopic property of an object related to kinetic energy of its particles. If we bring two objects of different temperatures into contact with each other, there is an energy flow (heat) between them. On average the faster-moving particles of the hotter object lose some of their kinetic energy when they collide with the slower particles of the colder object. The temperature changes for each object. The hotter object's temperature drops and the colder object's temperature rises.

A challenge arises for science teachers because everyday discussions of heat and temperature often include use of the phrases *heating things up* and *the amount of heat in an object*. Science teachers navigate this bridge between everyday language and scientific terminology to help students understand concepts of temperature and heat, but they must be careful not to introduce terms too early.

The National Science Education Standards use the term *heat* in the content standards related to physical science in grades K–12 to refer to internal, or thermal, energy. The Bench-

marks use the term *heat energy* in learning goals related to energy transformations for grades K–12 when referring to the same concept. In this book, we use the term *heat energy* in grades K–4 and *thermal energy* in grades 5–12 when discussing thermal energy. Teachers might consider staying with the term *heat energy* in elementary school discussions of this topic and introducing the term *thermal energy* in middle and high school as students move to study and understanding of the atomic nature of matter.

Related Ideas in *National Science Education Standards* (NRC 1996)

K–4 Properties of Objects and Materials

- Materials can exist in different states—solid, liquid, and gas. Some common materials, such as water, can be changed from one state to another by heating or cooling.

5–8 Properties and Changes in Properties of Matter

★ A substance has characteristic properties, such as density, a boiling point, and solubility, all of which are independent of the amount of sample. A mixture of substances often can be separated into the original substances using one or more of the characteristic properties.

9–12 Structure and Properties of Matter

- Solids, liquids, and gases differ in the distances and angles between molecules or atoms and therefore the energy that binds them together. In solids the structure is

---

★ Indicates a strong match between the ideas elicited by the probe and a national standard's learning goal.

nearly rigid; in liquids molecules or atoms move around each other but do not move apart; and in gases molecules or atoms move almost independently of each other and are mostly far apart.

9–12 Conservation of Energy and the Increase in Disorder

★ Heat consists of random motion and the vibrations of atoms, molecules, and ions. The higher the temperature, the greater the atomic or molecular motion.

Related Ideas in *Benchmarks for Science Literacy* (AAAS 1993)

K–2 Constancy and Change

★ Things change in some ways and stay the same in some ways.

3–5 Structure of Matter

• Heating and cooling cause changes in the properties of materials. Many kinds of changes occur faster under hotter conditions.

3–5 Constancy and Change

★ Some features of things may stay the same even when other features change.

6–8 Structure of Matter

• Atoms and molecules are perpetually in motion. Increased temperature means greater average energy of motion, so most substances expand when heated. In solids, the atoms are closely locked in position and can only vibrate. In liquids, the atoms or molecules have higher energy, are more

loosely connected, and can slide past one another; some molecules may get enough energy to escape into a gas. In gases, the atoms or molecules have still more energy and are free of one another except during occasional collisions.

9–12 Transformations of Energy

• Thermal energy in a material consists of the disordered motion of its atoms or molecules.

Related Research

• Many students think that the boiling point of water increases when the setting on a stove is "turned up." Much of this confusion is related to a misconception that heat and temperature are the same thing. Thus they are apt to argue that if you increase the amount of heat you will increase the boiling temperature (Driver et al. 1994).

• Students do not distinguish well between heat and temperature, and often believe that temperature is the measure of heat (AAAS 1993).

• Students may use the intuitive rule "more A, more B" or its converse "less A, less B" to reason what happens when you turn down the dial of a stove and the water still boils gently. Using this rule, students may think that when the temperature dial is turned down the boiling temperature is decreased (Stavy and Tirosh 2000).

Suggestions for Instruction and Assessment

• This probe can be followed up with an inquiry-based investigation. Ask the question, encourage students to commit to a prediction, then test it. The dissonance involved in discovering that the boiling temperature did not change should be followed with opportunities for students to discuss their ideas and resolve the dissonance.

- Have students use phase change graphs to analyze patterns and notice that when two phases are present (e.g., boiling water includes water in the liquid form and vapor) the temperature remains the same. Students can then use their graphs to predict what would need to happen next to increase the average kinetic energy and thus the temperature of the water particles.

- Connect the idea of a constant boiling point to other characteristic properties that remain constant under ordinary conditions, such as melting point, solubility, and density.

- Be sure to explicitly develop the generalization that boiling point is a constant property for all liquid substances, not just water.

- Do not introduce the difference between heat and temperature in the context of this probe until students are ready to understand this difference. Up through early middle school, it may suffice to keep this idea at an observational level and hold off on explanations until students are ready.

Related NSTA Science Store Publications and NSTA Journal Articles

American Association for the Advancement of Science (AAAS). 1993. *Benchmarks for science literacy.* New York: Oxford University Press.

Cavallo, A.M., and P. Dunphey. 2002. Sticking together: A learning cycle investigation about water. *The Science Teacher* 11 (Nov.): 24–28.

Driver, R., A. Squires, P. Rushworth, and V. Wood-Robinson. 1994. *Making sense of secondary science: Research into children's ideas.* London: RoutledgeFalmer.

Keeley, P. 2005. *Science curriculum topic study: Bridging the gap between standards and practice.* Thousand Oaks, CA: Corwin Press.

Purvis, D. 2006. Fun with phase changes. *Science & Children* 32 (5): 23–25.

National Research Council (NRC). 1996. *National science education standards.* Washington, DC: National Academy Press.

Robertson, W. 2002. *Energy: Stop faking it! Finally understanding science so you can teach it.* Arlington, VA: NSTA Press.

Related Curriculum Topic Study Guides
(Keeley 2005)
"Heat and Temperature"
"Physical Properties and Change"

References

American Association for the Advancement of Science (AAAS). 1993. *Benchmarks for science literacy.* New York: Oxford University Press.

Driver, R., A. Squires, P. Rushworth, and V. Wood-Robinson. 1994. *Making sense of secondary science: Research into children's ideas.* London: RoutledgeFalmer.

Keeley, P. 2005. *Science curriculum topic study: Bridging the gap between standards and practice.* Thousand Oaks, CA: Corwin Press.

National Research Council (NRC). 1996. *National science education standards.* Washington, DC: National Academy Press.

Stavy, R., and D. Tirosh. 2000. *How students (mis-) understand science and mathematics: Intuitive rules.* New York: Teachers College Press.

Boiling Time and Temperature

Ernesto is heating a pure liquid on a stove. He records the temperature a minute after the liquid starts to boil. After 20 minutes of boiling, he records the temperature again. When Ernesto compares the first temperature with the second, what do you think he will find? Circle your prediction.

A The boiling temperature did not change.

B The boiling temperature decreased.

C The boiling temperature increased.

Explain your thinking. Describe the "rule" or reasoning you used to make your prediction.

_____

_____

_____

_____

_____

_____

_____

_____

_____

Boiling Time and Temperature

Teacher Notes

Purpose

The purpose of this assessment probe is to elicit students' ideas about the characteristic property of boiling point. The probe is used to find out whether students recognize that the temperature of a boiling liquid stays constant no matter how long heat is applied.

Related Concepts

boiling and boiling point, change in state, characteristic properties, energy, heat, intensive properties of matter, temperature

Explanation

The best response is A: The boiling temperature did not change. Temperature is a measure of the average kinetic energy of the molecules in a system. When heat is applied to a liquid, the kinetic energy of the molecules increases. The motion of the liquid's molecules, and hence temperature of the liquid, increases until the temperature of the liquid reaches its boiling point (each pure substance has a specific temperature at which it will boil). Once a liquid is at its boiling point, the heat supplied to the system is used to overcome attractive forces between particles in the liquid. This results in the conversion of molecules from a liquid to a gas phase, allowing the molecules to escape into the air. The temperature of the remaining liquid essentially stays constant as heating continues. In the case of pure water, this is 100°C. (Impurities in tap water may result in a slight temperature rise during an extended period of boiling as the remaining solution becomes more concentrated.) The temperature will not rise again until all the water is transformed into a gaseous state. If the heat continues to be applied to the gaseous state, the temperature of the gas will rise.

Curricular and Instructional Considerations

Elementary Students

At the elementary level, students' experiences with materials are primarily observational. The idea of change is connected to properties of materials by subjecting them to temperature changes through heating and freezing. Upper elementary students are familiar with the changes in states of water from the solid to liquid to gas phase and vice versa. They learn how to use thermometers to measure the temperature of water. They may observe that water boils at 100°C (or slightly more if there are impurities in the tap water), but understanding the difference between heat and temperature and the notion of characteristic properties exceeds expectations for this grade level. However, the probe is useful in determining whether students intuitively or through their everyday experiences believe that increasing the amount of time a liquid is allowed to boil makes it "hotter" and thus increases its temperature.

Middle School Students

In middle school, students shift their focus from properties of materials to the characteristic properties of substances. Students are exposed to the characteristics of different states of matter and the properties associated with phase changes from solid to liquid to gas. Opportunities to observe and measure characteristic properties such as boiling and melting points can be used to distinguish and separate one substance from another. Students have had experiences with boiling liquids, and this probe may be useful in determining students' ideas about heat and temperature and whether they recognize that boiling temperature is a constant. Even though students can identify the boiling point of pure water as 100°C under normal conditions, they may still intuitively believe that the temperature rises the longer heat is applied to a boiling liquid.

High School Students

During high school, instructional opportunities connect the macroscopic properties of substances studied in middle school to microscopic properties. This probe may be useful in determining if students can relate the particulate nature of liquids and gases to the role of thermal energy in phase changes. It is a grade-level expectation that students can distinguish between heat and temperature, although it is still a very difficult concept for most students. They should be able to explain why pure boiling water stays at the same temperature regardless of how long it is boiling. However, if their preconceptions remain unchallenged throughout instruction, they may continue to believe that the longer water boils, the hotter it gets.

Administering the Probe

It may be helpful to have visual props for this probe. Make sure students understand the difference between a pure substance and an impure one (e.g., distilled water and saltwater). Students may want to know what type of liquid is boiling. If you use water, make sure students realize that the probe is asking about any pure liquid. Bring a beaker of water to a full boil. Continue to heat the boiling water as students respond to the probe and explain their thinking. Teachers may want to continue to probe ideas about heat, energy, temperature, boiling point, and states of matter by using this probe in conjunction with the probe "Turning the Dial" (p. 47) or "What's in the Bubbles?" (p. 65).

Related Ideas in *National Science Education Standards* (NRC 1996)

. . . .

K–4 Properties of Objects and Materials

- Materials can exist in different states— solid, liquid, and gas. Some common materials, such as water, can be changed from one state to another by heating or cooling.

5–8 Properties and Changes in Properties of Matter

★ A substance has characteristic properties, such as density, a boiling point, and solubility, all of which are independent of the amount of the sample. A mixture of substances often can be separated into the original substances using one or more of the characteristic properties.

9–12 Structure and Properties of Matter

- Solids, liquids, and gases differ in the distances and angles between molecules or atoms and therefore the energy that binds them together. In solids the structure is nearly rigid; in liquids molecules or atoms move around each other but do not move apart; and in gases molecules or atoms move almost independently of each other and are mostly far apart.

9–12 Conservation of Energy and the Increase in Disorder

★ Heat consists of random motion and the vibrations of atoms, molecules, and ions. The higher the temperature, the greater the atomic or molecular motion.

Related Ideas in *Benchmarks for Science Literacy* (AAAS 1993)

. . . .

3–5 Structure of Matter

- Heating and cooling cause changes in the properties of materials. Many kinds of changes occur faster under hotter conditions.

6–8 Structure of Matter

- Atoms and molecules are perpetually in motion. Increased temperature means greater average energy of motion, so most

---

★ Indicates a strong match between the ideas elicited by the probe and a national standard's learning goal.

substances expand when heated. In solids, the atoms are closely locked in position and can only vibrate. In liquids, the atoms or molecules have higher energy, are more loosely connected, and can slide past one another; some molecules may get enough energy to escape into a gas. In gases, the atoms or molecules have still more energy and are free of one another except during occasional collisions.

9–12 Transformations of Energy

- Thermal energy in a material consists of the disordered motion of its atoms or molecules.

Related Research

- Some students think that the boiling point of water increases the longer it is allowed to boil. Much of this confusion is related to a misconception that heat and temperature are the same thing. Thus, students are apt to argue that the longer you heat something, the hotter it gets (Driver et al. 1994).

- Students do not make a clear distinction between heat and temperature, and they often believe that temperature is the measure of heat (AAAS 1993).

- Students often will use a "more A, more B" type of intuitive rule for reasoning about what happens to the temperature the longer a liquid boils. Based on the everyday experience that the temperature of an object rises when heated, students may reason that the longer you heat a substance

after the onset of boiling, the higher the temperature will be (Stavy and Tirosh 2000).

- A standard laboratory exercise is to plot a time-temperature graph of water as it changes from melting ice to boiling water. Although students can readily see the steady temperature as they make their observations, the counterintuitiveness of the phenomenon often results in disbelief statements such as "this thermometer is not working properly" (Erickson and Tiberghien 1985, p. 64).

Suggestions for Instruction and Assessment

- This probe can be followed up with an inquiry-based investigation. Ask the question, encourage students to commit to a prediction, then test it. The dissonance involved in discovering that the boiling temperature did not change should be followed with opportunities for students to discuss their ideas and resolve the dissonance. However, be aware that tap water that contains impurities may change a little during 20 minutes of boiling. As water boils away, the remaining solution becomes more concentrated and boiling temperature increases slightly.

- Conduct a similar investigation to examine the effect of the continuous application of heat on the temperature of a substance existing in two different phases, such as a container filled with water containing ice cubes, snow, an ice "slush," or another familiar substance as it melts. Measure temperature at different time intervals while it is still melting. Contrast the findings from melting (a liquid and solid phase being heated over time) with boiling (a liquid and gas phase being heated over time).

- When having students develop time-temperature graphs, be aware that they may be able to explain their findings as shown on the graph yet revert to their belief that the temperature does not remain constant. Be sure to provide sufficient time to discuss the graphs and what they show.

- Be explicit about developing the generalization that a constant, specific boiling point applies to all liquid substances, not just water.

Related NSTA Science Store Publications and NSTA Journal Articles

American Association for the Advancement of Science (AAAS). 1993. *Benchmarks for science literacy.* New York: Oxford University Press.

Cavallo, A.M., and P. Dunphey. Sticking together: A learning cycle investigation about water. *The Science Teacher* 11 (Nov.): 24–28.

Driver, R., A. Squires, P. Rushworth, and V. Wood-Robinson. 1994. *Making sense of secondary science: Research into children's ideas.* London: RoutledgeFalmer.

Keeley, P. 2005. *Science curriculum topic study: Bridging the gap between standards and practice.* Thousand Oaks, CA: Corwin Press.

National Research Council (NRC). 1996. *National science education standards.* Washington, DC:

National Academy Press.

Purvis, D. 2006. Fun with phase changes. *Science & Children* 32 (5): 23–25.

Robertson, W. 2002. *Energy: Stop faking it! Finally understanding science so you can teach it.* Arlington, VA: NSTA Press.

Related Curriculum Topic Study Guides
(Keeley 2005)
"Physical Properties and Change"
"States of Matter"

References

American Association for the Advancement of Science (AAAS). 1993. *Benchmarks for science literacy.* New York: Oxford University Press.

Driver, R., A. Squires, P. Rushworth, and V. Wood-Robinson. 1994. *Making sense of secondary science: Research into children's ideas.* London: RoutledgeFalmer.

Erickson, G., and A. Tiberghien. 1985. Heat and temperature. In *Children's ideas in science,* eds. R. Driver, E. Guesne, and A. Tiberghien, 52–84. Milton Keynes, England: Open University Press.

Keeley, P. 2005. *Science curriculum topic study: Bridging the gap between standards and practice.* Thousand Oaks, CA: Corwin Press.

National Research Council (NRC). 1996. *National science education standards.* Washington, DC: National Academy Press.

Stavy, R., and D. Tirosh. 2000. *How students (mis-) understand science and mathematics: Intuitive rules.* New York: Teachers College Press.

Freezing Ice

Mia and Devon are having a summer party. They need to make two sizes of ice. The large blocks of ice will be put in a cooler to keep the cans of soda cold. The small ice cubes will keep the sodas in the glasses cold. They wondered how the temperature at which ice freezes is affected by size.

What do you think? Circle the answer that best matches your thinking.

A Small ice cubes freeze at a lower temperature than large blocks of ice.

B Small ice cubes freeze at a higher temperature than large blocks of ice.

C Small ice cubes and large blocks of ice freeze at the same temperature.

Explain your thinking. Describe the "rule" or reasoning you used for your answer.

_____

_____

_____

_____

_____

_____

_____

_____

_____

Freezing Ice

Teacher Notes

Purpose

The purpose of this assessment probe is to elicit students' ideas about freezing point. The probe is designed to find out whether students recognize that water freezes at the same temperature independent of the volume of water.

Related Concepts

characteristic properties, energy, freezing point, intensive properties of matter, temperature

Explanation

The best response is C: Small ice cubes and large blocks of ice freeze at the same temperature. The temperature at which pure water be-

gins to turn to ice, its freezing point, is 0°C. This temperature is the same regardless of how much water is being frozen. Freezing point is a characteristic property of matter that is independent of the amount of matter. Each liquid pure substance has a specific freezing point under standard conditions.

Curricular and Instructional Considerations

Elementary Students

At the elementary level, students' experiences with the properties of materials are primarily observational. The idea of change is connected to physical properties by subjecting materials to heating and cooling and observing what happens. Upper elementary students are familiar with the change in states of water from the solid to liquid to gas phase and vice versa. They should know how to use thermometers to measure the temperature of water. They may observe that water freezes at 0°C, but knowing the difference between heat and temperature is too complex for this age. The probe is useful in determining whether students intuitively think that the more water there is to freeze, the lower the temperature required to freeze it. They may also confuse the longer time it takes to freeze a larger volume of water with lowering the freezing temperature.

Middle School Students

In middle school, students shift their focus from properties of materials to the characteristic properties of the substances from which the materials are made. Students learn about the characteristics of different states of matter and the properties associated with phase changes from liquid to solid. Students in northern climates may draw on their everyday experience to connect the idea of freezing point to weather-related phenomena, knowing that icy conditions happen when the temperature reaches 0°C or 32°F. Understanding of characteristic properties, such as freezing point, is a grade-level expectation in the national standards. Students begin to understand the idea that pure substances have a specific freezing point and that this freezing point does not change under ordinary conditions. They are beginning to connect

the ideas of heat and temperature, although the idea of a constant temperature when two phases are present during a change in state is still counterintuitive to them.

High School Students

During high school, instructional opportunities connect the macroscopic properties of substances studied in middle school to microscopic properties. This probe may be useful in determining if students can relate the particulate nature of liquids and solids to the role of heat energy in phase changes. Students should be able to explain why water stays at the same temperature as it freezes regardless of how much water is in the sample. It is a grade-level expectation in the national standards that students can distinguish between heat and temperature, although this is difficult for most students to understand.

Administering the Probe

You may wish to use visual props for this probe, such as a small tray of ice cubes and a block of frozen ice. To continue to probe reasoning about heat, energy, temperature, and states of matter, you can combine this probe with the "Turning the Dial" (p. 47) and "Boiling Time and Temperature" (p. 53) probes.

In explaining their reasoning, some students tend to focus on the time it takes the ice to freeze, rather than the temperature. You may need to remind them that the probe is asking for an explanation of how size affects the temperature, not how long it takes the ice to freeze.

Related Ideas in *National Science Education Standards* (NRC 1996)

K–4 Properties of Objects and Materials

• Materials can exist in different states—solid, liquid, and gas. Some common materials, such as water, can be changed from one state to another by heating or cooling.

5–8 Properties and Changes in Properties of Matter

★ A substance has characteristic properties, such as density, a boiling point, and solubility, all of which are independent of the amount of the sample. A mixture of substances often can be separated into the original substances using one or more of the characteristic properties.

9–12 Structure and Properties of Matter

• Solids, liquids, and gases differ in the distances and angles between molecules or atoms and therefore the energy that binds them together. In solids the structure is nearly rigid; in liquids molecules or atoms move around each other but do not move apart; and in gases molecules or atoms move almost independently of each other and are mostly far apart.

9–12 Conservation of Energy and the Increase in Disorder

★ Heat consists of random motion and the vibrations of atoms, molecules, and ions.

The higher the temperature, the greater the atomic or molecular motion.

Related Ideas in *Benchmarks for Science Literacy* (AAAS 1993)

K–2 Constancy and Change

★ Things change in some ways and stay the same in some ways.

3–5 Constancy and Change

★ Some features of things may stay the same even when other features change.

• Heating and cooling cause changes in the properties of materials.

6–8 Structure of Matter

• Atoms and molecules are perpetually in motion. Increased temperature means greater average energy of motion, so most substances expand when heated. In solids, the atoms are closely locked in position and can only vibrate. In liquids, the atoms or molecules have higher energy, are more loosely connected, and can slide past one another; some molecules may get enough energy to escape into a gas. In gases, the atoms or molecules have still more energy and are free of one another except during occasional collisions.

Related Research

• Students may use the intuitive rule "more A, more B" or its converse "less A, less B" to reason what happens when you freeze different

---

★ Indicates a strong match between the ideas elicited by the probe and a national standard's learning goal.

volumes of water. Using this rule, students may think that when there is more water to freeze, the freezing temperature needs to be lower (Stavy and Tirosh 2000).

- Up to the age of 12 students are familiar with the term *temperature* and are able to use a thermometer to measure the temperature of objects or materials, but they actually have a fairly limited concept of the term. They rarely use *temperature* to spontaneously describe the condition of an object (Erickson and Tiberghien 1985).

- In certain experimental situations, many students believe that the temperature of an object is related to its size. In one study more than 50% of 12-year-old students thought that "a larger ice cube would have a colder temperature and hence the larger ice cube would take longer to melt" (Erickson and Tiberghien 1985, p. 61).

- Cosgrove and Osborne (1980) interviewed students about their ideas related to change in state and noticed that students generally do not regard a change in state as being related to a specific temperature (Driver et al. 1994).

Suggestions for Instruction and Assessment

- This probe can be followed up with an inquiry-based investigation. Ask the question, encourage students to commit to a prediction, then test it by placing a thermometer in a small ice cube tray and a large container of water and reading the thermometer as the ice begins to form. The

dissonance involved in discovering that the freezing temperature did not change should be followed with opportunities for students to discuss their ideas and resolve the dissonance.

- Instead of having students graph the change in state from liquid/gas to liquid to liquid/solid to solid, have them start with collecting part of the data and making a prediction based on the pattern observed. Use two different volumes of water to collect and compare data. Begin by graphing two different volumes of boiling water and have students notice that when two phases are present (liquid and gas state) the temperature remains the same, regardless of sample size. Allow the boiling water to cool and observe that the temperature steadily decreases when one phase is present (liquid), although the rate may differ in the two samples. Students can use their graphs to predict what would happen to the temperature of two different volumes once the water begins to freeze and two phases (liquid water and ice) are present, and they can then test their predictions. This investigation allows students to not only observe the patterns during change in state but also notice that there are constants in the plateaus of these patterns regardless of the volume of sample.

- Explicitly connect the idea of a specific freezing point to a specific boiling point and other characteristic properties in order to develop the generalization that characteristic properties are independent of the

amount of a sample.

- Compare freezing point to melting point to show that when two phases are present, the temperature is the same regardless of whether you start by melting ice or start by freezing water.
- Be explicit about developing the generalization that freezing point is specific under standard conditions for all pure liquid substances, not just water.

Related NSTA Science Store Publications and NSTA Journal Articles

American Association for the Advancement of Science (AAAS). 1993. *Benchmarks for science literacy.* New York: Oxford University Press.

Driver, R., A. Squires, P. Rushworth, and V. Wood-Robinson. 1994. *Making sense of secondary science: Research into children's ideas.* London: RoutledgeFalmer.

Keeley, P. 2005. *Science curriculum topic study: Bridging the gap between standards and practice.* Thousand Oaks, CA: Corwin Press.

Link, L., and E. Christmann. 2004. A different phase change. *Science Scope* 28 (3): 52–54.

National Research Council (NRC). 1996. *National science education standards.* Washington, DC: National Academy Press.

Purvis, D. 2006. Fun with phase changes. *Science & Children* 32 (5): 23–25.

Robertson, W. 2002. *Energy: Stop faking it! Finally understanding science so you can teach it.* Arlington, VA: NSTA Press.

Related Curriculum Topic Study Guides

(Keeley 2005)

"Heat and Temperature"

"Physical Properties and Change"

References

American Association for the Advancement of Science (AAAS). 1993. *Benchmarks for science literacy.* New York: Oxford University Press.

Cosgrove, M., and R. Osborne. 1980. *Physical change.* LISP Working Paper 26. Hamilton, New Zealand: University of Waikato, Science Education Research Unit.

Driver, R., A. Squires, P. Rushworth, and V. Wood-Robinson. 1994. *Making sense of secondary science: Research into children's ideas.* London: RoutledgeFalmer.

Erickson, G., and A. Tiberghien. 1985. Heat and temperature. In *Children's ideas in science,* eds. R. Driver, E. Guesne, and A. Tiberghien, 52–84. Milton Keynes, England: Open University Press.

Keeley, P. 2005. *Science curriculum topic study: Bridging the gap between standards and practice.* Thousand Oaks, CA: Corwin Press.

National Research Council (NRC). 1996. *National science education standards.* Washington, DC: National Academy Press.

Stavy, R., and D. Tirosh. 2000. *How students (mis-) understand science and mathematics: Intuitive rules.* New York: Teachers College Press.

What's in the Bubbles?

Hannah is boiling water in a glass tea kettle. She notices bubbles forming on the bottom of the kettle that rise to the top and wonders what is in the bubbles. She asks her family what they think, and this is what they say:

Dad: "They are bubbles of heat."

Calvin: "The bubbles are filled with air."

Grandma: "The bubbles are an invisible form of water."

Mom: "The bubbles are empty—there is nothing inside them."

Lucy: "The bubbles contain oxygen and hydrogen that separated from the water."

Which person do you most agree with and why? Explain your thinking.

_____

_____

_____

_____

_____

_____

_____

_____

_____

_____

What's in the Bubbles?

Teacher Notes

Purpose

The purpose of this assessment probe is to elicit students' ideas about particles during a change in state. The probe is designed to find out if students recognize that the bubbles formed when water boils are the result of liquid water changing into water vapor.

Related Concepts

atoms or molecules, boiling and boiling point, change in state, energy

Explanation

The best response is Grandma's: The bubbles are an invisible form of water. This invisible water is called water vapor, a gaseous form of water that is not visible; it is unlike steam, which contains some condensed liquid water. When water is heated, the energy supplied to the system results in an increase in molecular motion. If enough heat is supplied, the molecules have so much energy that they can no longer remain loosely connected, sliding past one another as they do in a liquid. The energy now allows the attractive forces between water molecules to be overcome, and they form an "invisible" gas in the form of water vapor. Since the molecules in the gas phase are so much farther apart than in the liquid phase, they have a much lower density, are more buoyant (causing them to "bubble up"), and escape into the air. The bubble is the invisible water vapor.

Curricular and Instructional Considerations

Elementary Students

At the elementary level, students have experiences observing changes in state. The idea of change is connected to physical properties of materials by subjecting materials to heating and freezing. Water is often used as a familiar material for observing phase changes. Elementary students know change in states of water from the solid to liquid to gas phase, although the change from liquid to gas phase is a more abstract idea developed more fully in upper elementary grades.

In early elementary grades, students' experience with bubbles that result when water boils is primarily observational and is often linked to experiences at home boiling water on a stove. It is too early to introduce the abstract idea of invisible water molecules that make up water vapor. However, students can develop the precursor idea that water, in the form of invisible water vapor, escapes from the surface of an uncovered liquid. It may be too soon to introduce the idea that bubbles of boiling water contain water vapor, although students can observe steam going into the air from water that boils, even though steam contains some tiny droplets of water. Students must understand the simpler idea that water goes into the air in a form we cannot see before the idea of kinetic molecular theory, which helps explain why bubbles form and what they are, is introduced in middle school. The notion that water vapor is a gas is a grade-

level expectation in the national standards. Children develop conceptions about bubbles early on through their everyday experiences, so it is not too early to ask students their ideas about boiling and bubbles. However, it is best to hold off on expecting a scientific explanation until students are ready.

Middle School Students

In middle school, students have opportunities to examine the characteristics of different states of matter, and they begin to conceptualize the particle movements associated with phase changes from solid to liquid to gas. Students observe and measure characteristic properties such as boiling point and melting point. Students have had varied experiences with boiling water. They compare evaporation of a liquid under ordinary ambient conditions as well as in situations where increased application of heat is involved, such as boiling water. This probe is useful in determining students' preconceptions related to the common phenomenon of bubbles forming in boiling water.

High School Students

During high school, instructional opportunities connect the macroscopic properties of substances studied in middle school to a microscopic level. An understanding of kinetic molecular theory is a grade-level expectation in the standards that can be used to explain what the bubbles in boiling water are. This probe may be useful in determining if students revert to their earlier preconceptions about bubbles or if they can explain what is happening at a molecular level.

Administering the Probe

You may wish to use visual props for this probe. Bring a beaker of water or some other clear glass, boiling-safe container to a full boil so that students can see the bubbles forming and rising to the surface. Be sure students are wearing safety goggles and are not too close to the heat source if they are observing the boiling up close. Continue to heat the boiling water as students respond to the probe and explain their thinking. Teachers may want to continue to probe students' ideas about boiling by combining this probe with the "Turning the Dial" (p. 47) and "Boiling Time and Temperature" (p. 53) probes.

Related Ideas in *National Science Education Standards* (NRC 1996)

K–4 Properties of Objects and Materials

• Materials can exist in different states—solid, liquid, and gas. Some common materials, such as water, can be changed from one state to another by heating or cooling.

5–8 Properties and Changes in Properties of Matter

• A substance has characteristic properties, such as density, a boiling point, and solubility, all of which are independent of the amount of the sample. A mixture of substances often can be separated into the

original substances using one or more of the characteristic properties.

9–12 Structure and Properties of Matter

★ Solids, liquids, and gases differ in the distances and angles between molecules or atoms and therefore the energy that binds them together. In solids the structure is nearly rigid; in liquids molecules or atoms move around each other but do not move apart; and in gases molecules or atoms move almost independently of each other and are mostly far apart.

Related Ideas in *Benchmarks for Science Literacy* (AAAS 1993)

3–5 Structure of Matter

• Heating and cooling cause changes in the properties of materials. Many kinds of changes occur faster under hotter conditions.

3–5 The Earth

★ When liquid water disappears, it turns into a gas (vapor) in the air and can reappear as a liquid when cooled, or as a solid if cooled below the freezing point of water. Clouds and fog are made up of tiny droplets of water.

6–8 Structure of Matter

★ Atoms and molecules are perpetually in motion. Increased temperature means greater average energy of motion, so most substances expand when heated. In solids, the atoms

---

★ Indicates a strong match between the ideas elicited by the probe and a national standard's learning goal.

are closely locked in position and can only vibrate. In liquids, the atoms or molecules have higher energy, are more loosely connected, and can slide past one another; some molecules may get enough energy to escape into a gas. In gases, the atoms or molecules have still more energy and are free of one another except during occasional collisions.

Related Research

- In a study by Barker (2004), many students ages 8–17 thought that the bubbles seen in boiling water are made of heat, air, oxygen, or hydrogen. Another conception was a change in state model that involved molecules breaking up on boiling and reforming on condensing. Barker also discovered that students find it hard to appreciate the reversibility of phase changes, thinking of each process as a separate event.

- Students' understanding of boiling precedes their understanding of evaporation from surfaces such as dishes and roads. In a sample of students ages 6–8, 70% understood that when water boils vapor comes from it and that the vapor is made of water. However, the same students did not recognize that when a wet surface dries, the water turns to water vapor (Driver et al. 1994).

Suggestions for Instruction and Assessment

- Use the phenomenon of bubbles to explain what happens to water molecules during a change in state from boiling liquid to gas.

- Encourage students to draw the stages of what they think is happening to the water as it is heated. Continue drawing right up to the stage where bubbles are formed and rising to the top and bursting. Carefully note how students get to the bubble stage—do the bubbles appear spontaneously in their drawings, or does the act of drawing help them make sense of what is happening to the water to form bubbles?

- Students may have trouble accepting that water vapor is in the bubbles if they do not understand the idea that water vapor is invisible. Help students contrast the concept of invisible water vapor with visible water in the air such as clouds and fog, which are made of tiny suspended droplets rather than water molecules spread far apart.

- Ask students to observe and describe what happens to the water level as the water boils. Encourage them to explain where the water went. How was it able to leave the glass container? Probe students to consider how the bubbles were involved in decreasing the water level. Challenge students who had the idea that the bubbles were air or nothing to explain how their model could account for the decreased water level.

- Consider how to present phase changes as reversible. Allow students to see heating and cooling cycles for themselves, so they can realize that phase changes do not result in a new substance being formed. This cycle may help them see that the water escapes as a gas in the bubbles and can be recovered again through cooling.

- By upper elementary grades, students should begin using terminology such as *water vapor*. Using the correct terminology and developing an understanding that water is in the air may help them overcome the idea that water changes into air rather than remaining the same substance but in a form that you cannot see.

- Be cautious when using the term *steam* with students to describe the gas or vapor form of water. What students are actually seeing over the boiling water when they refer to steam is a wispy mist—it is visible because it is water in a gaseous state that also contains tiny water droplets. Those tiny droplets scatter light at their surfaces, allowing us to "see" the "steam" in much the same way that we can see fog or clouds. The common use of the word steam is different from the way scientists or engineers use the word *steam*. To them, steam and vapor are both invisible forms of water in the gaseous state. However, when students (and often teachers) use the word *steam* in the context of this probe, they are usually calling the visible substance that forms above the boiling water a gas. Technically this common use of the word *steam* is incorrect since a gas is invisible. The Standards use the term *vapor* (not steam) to describe the invisible, gaseous form of water and explicitly point out that clouds and fog are made up of tiny droplets of water in order to distinguish forms of water in the air that we can see from forms we cannot see. Older students may be introduced to the scientific use of the word

steam and compare it to how it is commonly used in our everyday language, once they have grasped the idea that substances in the gaseous state are not visible.

Related NSTA Science Store Publications and NSTA Journal Articles

See articles and publications listed on page 58.

Related Curriculum Topic Study Guides
(Keeley 2005)
"Physical Properties and Change"
"States of Matter"

References

American Association for the Advancement of Science (AAAS). 1993. *Benchmarks for science literacy.* New York: Oxford University Press.

Barker, V. 2004. *Beyond appearances: Students' misconceptions about basic chemical ideas. A report prepared for the Royal Society of Chemistry.* Cambridge, England.

Driver, R., A. Squires, P. Rushworth, and V. Wood-Robinson. 1994. *Making sense of secondary science: Research into children's ideas.* London: RoutledgeFalmer.

Keeley, P. 2005. *Science curriculum topic study: Bridging the gap between standards and practice.* Thousand Oaks, CA: Corwin Press.

National Research Council (NRC). 1996. *National science education standards.* Washington, DC: National Academy Press.

Chemical Bonds

Three students were discussing their ideas about chemical bonds. This is what they said:

Janre: "I think a chemical bond is produced by a molecule. It is a substance made up of matter that holds atoms together."

Will: "I think a chemical bond is an attraction between atoms. It is not made up of matter."

Leta: "I think a chemical bond is a structural part of an atom that connects it to other atoms."

Which student do you most agree with and why? Explain your thinking.

_____

_____

_____

_____

_____

_____

_____

_____

_____

_____

_____

Chemical Bonds

Teacher Notes

Purpose

The purpose of this assessment probe is to elicit students' ideas about chemical bonds. The probe is designed to find out if students think bonds are physical matter or attractions between electrons.

Related Concepts

atoms or molecules, chemical bonds

Explanation

The best response is Will's. A chemical bond is an attraction between atoms; it is not made up of matter that holds atoms together, and it is not a structural part of an atom. Two or more atoms are linked together by chemical bonds. There are several types of chemical bonds, including covalent bonds, ionic bonds, metallic bonds, and hydrogen bonds. Chemical bonds are formed between atoms as a result of an attraction between their electrons. The bond exists as an attractive force between the atoms where electrons are transferred or shared.

Curricular and Instructional Considerations

Elementary Students

The concept of chemical bonds far exceeds the ideas expected of students at the elementary level. However, upper elementary students have often seen representations of molecules and may begin to form the "ball-and-stick" idea that there is a structure or "glue" holding particles together.

Middle School Students

In middle school, students develop the idea that atoms join together to form molecules or crystalline arrays. They encounter the term *chemical bond* in both life science and physical science and have a concept of atoms being joined together, but an understanding of the mechanism by which electrons are shared or transferred, resulting in an attraction that holds atoms together, exceeds the middle school level. Students at this level see a variety of representations of molecules and ionic substances, including ball-and-stick models, which may contribute to their conception of a physical chemical bond. They learn that models, such as physical and graphic models of molecules and crystal arrays, do not entirely represent the real thing. These models can be used without going into details about the formation or types of chemical bonds.

High School Students

Students at this level develop a deeper understanding of the microscopic nature of molecules, atoms, and parts of atoms, including the types of chemical bonds formed by the interaction of electrons. The nature of the atom, including its interaction with other atoms, is still an abstract, difficult idea for many students. Because representations of molecules and compounds, including physical models and symbolic drawings, are commonly used in high school science, it is important to take the time to determine whether students have a conception of a chemical bond as a physical entity or a force of attraction. Many students can define the types of chemical bonds and the mechanism in which atoms are joined together yet still harbor the common misconception that a chemical bond is a structural component of a substance or a glue-like form of matter.

Administering the Probe

This probe is appropriate for middle school and high school students, although middle school students should not be expected to know the detailed mechanism of bonding.

Related Ideas in *National Science Education Standards* (NRC 1996)

9–12 Structure and Properties of Matter

• Atoms interact with one another by transferring or sharing electrons that are farthest from the nucleus. The outer electrons govern the chemical properties of the element.

★ Bonds between atoms are created when electrons are paired up by being transferred or shared. A substance composed of a single kind of atom is called an element. The atoms may be bonded together into molecules or crystalline solids. A compound is formed when two or more kinds of atoms bind together chemically.

Related Ideas in *Benchmarks for Science Literacy* (AAAS 1993)

3–5 Structure of Matter

• Materials may be composed of parts that are too small to be seen without magnification.

6–8 Structure of Matter

• All matter is made up of atoms, which are too small to see directly through a microscope. The atoms of any element are alike but are different from atoms of other elements. Atoms may stick together in well-defined molecules or may be packed together in large arrays. Different arrangements of atoms into groups compose all substances.

9–12 Structure of Matter

★ Atoms are made up of a positive nucleus surrounded by negative electrons. An atom's electron configuration, particularly the outermost electrons, determines how the atom can interact with other atoms. Atoms form bonds to other atoms by transferring or sharing electrons.

• Atoms often join with one another in vari-

ous combinations in distinct molecules or in repeating three-dimensional crystal patterns. An enormous variety of biological, chemical, and physical phenomena can be explained by changes in the arrangement and motion of atoms and molecules.

Related Research

• An extension of the idea that atoms "need" to form bonds is that atoms "make decisions" about forming bonds. This may come from analogies used in teaching such as holding hands or finding a new partner (Barker 2004).

• In general, students have difficulty in developing an adequate conception of the chemical combination of elements until they can interpret *combination* at the molecular level (Driver et al. 1994).

• Students have difficulty interpreting the use of ball-and-stick models for ionic lattices. Twenty-seven Australian 17-year-olds were interviewed in a study by Butts and Smith (1987) using a ball-and-stick model of sodium chloride. Students confused the six sticks around each ball as "one ionic and five physical bonds." Only two of the students mentioned that the sticks were used merely to hold the balls in place for the model (Driver et al. 1994).

Suggestions for Instruction and Assessment

• The use of anthropomorphic analogies to explain how bonds form should be avoided. These analogies give students false ideas

---

★ Indicates a strong match between the ideas elicited by the probe and a national standard's learning goal.

about atoms "wanting" to form bonds, "needing" a certain number of bonds, or "finding a partner." The analogies tend to confuse organisms' behavior with chemical behavior (Barker 2004).

- When using representations such as symbolic drawings of compounds and ball-and-stick models, explicitly state that the lines do not represent actual physical structures at the atomic and molecular level.

- Connect the idea of representations of chemical bonds to students' understanding of how models are used to represent structures and phenomena. Provide an opportunity for students to critique the representations of molecules and compounds, describing the limitations of representations in depicting an actual molecule or compound.

Related NSTA Science Store Publications and NSTA Journal Articles

American Association for the Advancement of Science (AAAS). 1993. *Benchmarks for science literacy.* New York: Oxford University Press.

Cobb, C., and M. Fetterolf. 2005. *The joy of chemistry: The amazing science of familiar things.* Amherst, NY: Prometheus Books.

Driver, R., A. Squires, P. Rushworth, and V. Wood-Robinson. 1994. *Making sense of secondary science: Research into children's ideas.* London: RoutledgeFalmer.

Keeley, P. 2005. *Science curriculum topic study: Bridging the gap between standards and practice.*

Thousand Oaks, CA: Corwin Press.

National Research Council (NRC). 1996. *National science education standards.* Washington, DC: National Academy Press.

Related Curriculum Topic Study Guide
(Keeley 2005)
"Chemical Bonding"

References

American Association for the Advancement of Science (AAAS). 1993. *Benchmarks for science literacy.* New York: Oxford University Press.

Barker, V. 2004. *Beyond appearances: Students' misconceptions about basic chemical ideas. A report prepared for the Royal Society of Chemistry.* Cambridge, UK.

Butts, B., and R. Smith. 1987. HSC chemistry students' understanding of the structure and properties of molecular and ionic compounds. *Research in Science Education* 17 (1): 192–201.

Driver, R., A. Squires, P. Rushworth, and V. Wood-Robinson. 1994. *Making sense of secondary science: Research into children's ideas.* London: RoutledgeFalmer.

Keeley, P. 2005. *Science curriculum topic study: Bridging the gap between standards and practice.* Thousand Oaks, CA: Corwin Press.

National Research Council (NRC). 1996. *National science education standards.* Washington, DC: National Academy Press.

Ice-Cold Lemonade

It was a hot summer day. Mattie poured herself a glass of lemonade. The lemonade was warm, so Mattie put some ice in the glass. After 10 minutes, Mattie noticed that the ice was melting and the lemonade was cold. Mattie wondered what made the lemonade get cold. She had three different ideas. Which idea do you think best explains why the lemonade got cold? Circle your answer.

A The coldness from the ice moved into the lemonade.

B The heat from the lemonade moved into the ice.

C The coldness and the heat moved back and forth until the lemonade cooled off.

Explain your thinking. Describe the "rule" or reasoning you used for your answer.

_____

_____

_____

_____

_____

_____

_____

_____

_____

Ice-Cold Lemonade

Teacher Notes

Purpose

The purpose of this assessment probe is to elicit students' ideas about the transfer of energy. The probe is designed to determine whether students recognize that heat flows from warmer objects or areas to cooler ones.

Related Concepts

conduction, energy, energy transfer, heat

Explanation

The best response is B: The heat from the lemonade moved into the ice. Heat (thermal energy) is associated with the motion of molecules in a substance. This energy is transferred from one place to another through the processes of

heat flow. This thermal energy will only move from a warmer object to a cooler object, never the other way around. In the case of the lemonade and ice, as the molecules in the warmer lemonade came in contact with the molecules in the cooler ice, heat flowed into the ice from the lemonade. This process cooled the lemonade and melted the ice.

Common language contains many references to the idea of "cold" moving from place to place. Children are advised to close a refrigerator door, so as not to "let the cold out," and we complain about winter chills that "get into your bones." Such phrases reinforce the common notion that something known as cold can move from place to place. Because what we sense as warm or cold simply refers to the average kinetic energy (thermal energy) of an object's molecules, these references to cold moving are generally misnomers for the transfer of heat energy from warmer to cooler objects.

Curricular and Instructional Considerations

Elementary Students

In the early grades students use the terms *heat, hot,* and *cold* to describe encounters with objects and their surroundings. They have experiences mixing same and different amounts of hot and cold water together and finding the resulting temperature. As they reach the age of eight or nine they can talk about heat as a type of continuum from cold to hot, but they commonly associate heat with objects such as the stove, the Sun, or a fire. Developing the formal idea of heat transfer is difficult for younger elementary students and can wait until middle school. At this grade level it is sufficient for students to know that heat moves from one place to another, which can be observed with their senses. This probe is useful in determining students' early conceptions of the flow of heat.

Middle School Students

Students at this level have a general concept of heat but still associate it more with the nature of objects rather than energy transfer. Students' experiences with the movement of heat expand to include conduction, convection, and radiation. By the end of middle school, students should be able to connect their understanding of the motion of molecules to the transfer of heat in examples involving conduction and convection. Even with formal instruction, middle school students will still have difficulty understanding the direction of heat flow. This probe is useful in determining whether students hold on to their preconceived notions, particularly the idea that cold moves out of an object or substance.

High School Students

The concepts of heat, thermal energy, and energy transfer can now be extended into new contexts, including nuclear reactions and biological energy transfers. However, it is still likely that students will hold onto their preconceptions about the movement of cold, possibly until the laws of thermodynamics are introduced in physics. This probe is useful in determining whether these ideas persist.

Administering the Probe

You may wish to use visual props for this probe. For example, pour a glass of warm lemonade. Place a thermometer in the glass of lemonade and tell the class what the temperature of the lemonade is. Add ice to the glass of lemonade. After 10 minutes, tell the class what the temperature of the iced lemonade is and pose the question in the probe. Be aware that the language in the probe is intentional. The word *moved* was used instead of *transferred* to avoid memorized definitions of energy transfer. You may want to ask students to draw their ideas, noting whether they perceive heat as a substance that moves, similar to the historical

"caloric" model, or use ideas about the motion of particulate matter.

Related Ideas in *National Science Education Standards* (NRC 1996)

K–4 Light, Heat, Electricity, and Magnetism

★ Heat can be produced in many ways, such as burning, rubbing, or mixing one substance with another. Heat can move from one object to another.

5–8 Transfer of Energy

• Energy is a property of many substances and is associated with heat, light, electricity, mechanical motion, sound nuclei, and the nature of a chemical. Energy is transferred in many ways.

★ Heat moves in predictable ways, flowing from warmer objects to cooler ones until both reach the same temperature.

• In most chemical and nuclear reactions, energy is transferred into or out of a system. Heat, light, mechanical motion, or electricity might all be involved in such transfers.

9–12 Conservation of Energy and the Increase of Disorder

• Heat consists of random motion and the vibrations of atoms, molecules, and ions. The higher the temperature, the greater the atomic or molecular motion.

★ Everything tends to become less organized and less orderly over time. Thus in all energy

transfers the overall effect is that the energy is spread out uniformly. Examples are the transfer of energy from hotter to cooler objects by conduction and convection and the warming of our surroundings when we burn fuel.

Related Ideas in *Benchmarks for Science Literacy* (AAAS 1993)

3–5 Energy Transformation

★ When warmer things are put with cooler ones, the warm ones lose heat and the cool ones gain it until they are all the same temperature. A warmer object can warm a cooler one by contact or at a distance.

6–8 Energy Transformation

• Thermal energy is transferred through a material by the collisions of atoms within the material. Thermal energy can also be transferred by means of currents in air, water, or other fluids.

• Energy appears in different forms and can be transformed within a system. Thermal energy is associated with the temperature of an object.

9–12 Transformations of Energy

• Thermal energy in a system is associated with the disordered motions of its atoms or molecules.

Related Research

• Middle school students often do not explain the process of heating and cooling in terms

---

★ Indicates a strong match between the ideas elicited by the probe and a national standard's learning goal.

of heat being transferred. When transfer ideas are involved, some students will think that cold is being transferred from a colder to warmer object. Other students think that both heat and cold are transferred at the same time (AAAS 1993).

- Middle and high school students do not always explain heat-exchange phenomena as interactions. For example, students may say that objects tend to cool down or release heat spontaneously without acknowledging that the object has come in contact with a cooler object or area (AAAS 1993).

- Numerous studies have shown that few middle and high school students understand the molecular basis of heat transfer after instruction. Difficulties in understanding remain even with instruction that is specially designed to explicitly address the difficulty of understanding heat transfer (AAAS 1993).

- Many researchers have found that children have difficulty understanding heat-related ideas. Harris (1981) and other researchers suggested that much of the confusion about heat comes from the words we use and that children tend to think of heat as a substance that flows from one place to another. Cold is also thought of as an entity like heat, with many children thinking that cold is the opposite of heat rather than being part of the same continuum (Driver, Guesne, and Tiberghien 1985; Driver et al. 1994).

Suggestions for Instruction and Assessment

- The *Benchmarks* (AAAS 1993, p. 81) state

that energy is a major exception to the principle that students should understand an idea before giving them a label for it. Because energy is such a mysterious concept, children can actually benefit from hearing the term and talking about it before being able to define it. Developing a formal conception of energy and energy-related concepts should wait until students are ready.

- Elementary students should have multiple experiences putting warmer and cooler things together, measuring the temperature, and describing the result. It is not until middle school that students begin to describe and draw a particulate model to explain what happens.

- In upper elementary grades, students can investigate warm and cold objects, observing how heat seems to spread from one area to another. Starting with objects that are warmer than their immediate environment to investigate heat transfer may make more sense than starting with objects that are colder than their surrounding environment.

- Computer probeware may be more effective than ordinary thermometers in helping students observe small changes in temperature during heat transfer.

- Be aware that many students think that cold moves. When developing the idea of heat moving from warmer to cooler areas, have students generate examples of everyday phrases that describe the movement of cold, such as "shut the door or you will let all the cold in." Engage students in critiquing these phrases in terms of how energy moves.

- Explicitly address the idea of interactions when teaching about energy transfer so that students do not develop the notion of it being one-sided. Have students identify the materials or objects involved in the interactions.

- Instruction on heat and heat transfer should be carried out over the long term and not done in one short unit. These are difficult and abstract ideas, and it takes time and multiple experiences for students to use these ideas scientifically.

- High school students should have multiple opportunities to use heat energy transfer ideas in multiple contexts, including chemical, nuclear, geologic, and biological contexts. Revisiting heat transfer ideas reinforces the concept and helps students see how powerful the "big idea" of energy transfer is in explaining a wide range of phenomena.

Related NSTA Science Store Publications and NSTA Journal Articles

American Association for the Advancement of Science (AAAS). 1993. *Benchmarks for science literacy.* New York: Oxford University Press.

Ashbrook, P. 2006. The matter of melting. *Science & Children* (Jan.): 18–21.

Driver, R., A. Squires, P. Rushworth, and V. Wood Robinson. 1994. *Making sense of secondary science: Research into children's ideas.* London. RoutledgeFalmer.

Keeley, P 2005. *Science curriculum topic study: Bridging the gap between standards and practice.* Thousand Oaks, CA: Corwin Press.

May, K., and M. Kurbin. 2003. To heat or not to heat. *Science Scope* (Feb.): 38–41.

National Research Council (NRC). 1996. *National science education standards.* Washington, DC: National Academy Press.

Robertson, W. 2002. *Energy: Stop faking it! Finally understanding science so you can teach it.* Arlington, VA: NSTA Press.

Related Curriculum Topic Study Guides

(Keeley 2005)
"Energy"
"Heat and Temperature"

References

American Association for the Advancement of Science (AAAS). 1993. *Benchmarks for science literacy.* New York: Oxford University Press.

Driver, R., A. Squires, P. Rushworth, and V. Wood-Robinson, 1994. *Making sense of secondary science: Research into children's ideas.* London: RoutledgeFalmer.

Driver, R., E. Guesne, and A. Tiberghien. 1985. *Children's ideas in science.* Milton Keynes, England: Open University Press.

Harris, W. F. 1981. Heat in undergraduate education, or isn't it time we abandoned the theory of caloric? *International Journal of Mechanical Engineering Education* 9: 317–325.

Keeley, P. 2005. *Science curriculum topic study: Bridging the gap between standards and practice.* Thousand Oaks, CA: Corwin Press.

National Research Council (NRC). 1996. *National science education standards.* Washington, DC: National Academy Press.

Mixing Water

Melinda filled two glasses of equal size half-full with water. The water in one glass was 50 degrees Celsius. The water in the other glass was 10 degrees Celsius. She poured one glass into the other, stirred the liquid, and measured the temperature of the full glass of water.

What do you think the temperature of the full glass of water will be after the water is mixed? Circle your prediction.

A 20 degrees Celsius

B 30 degrees Celsius

C 40 degrees Celsius

D 50 degrees Celsius

E 60 degrees Celsius

Explain your thinking. Describe the "rule" or reasoning you used for your answer.

_____

_____

_____

_____

_____

_____

_____

_____

_____

_____

Mixing Water

Teacher Notes

50°C   10°C

Purpose

The purpose of this assessment probe is to elicit students' ideas about temperature and energy transfer. The probe is designed to find out whether students recognize that a transfer of energy from the warm water to the cool water occurs until they reach the same temperature. Additionally, students' explanations reveal whether they use an addition, subtraction, or averaging strategy to determine the resulting temperature.

Related Concepts

conduction, energy, energy transfer, heat, temperature

Explanation

The best response is B: 30°C. (In actuality it would be slightly less, because a small amount of heat is transferred from the water to the glass and the surrounding environment in the process.) Temperature is a measure of the average motion of the particles that make up the water. The two separate samples of water are at different temperatures, meaning the average energy of the particles is less in the cooler (10°C) sample. When the cooler water and the warmer water are mixed together, a transfer of energy (conduction) occurs between particles when they come in contact with each other. The flow of heat moves from the molecules in the warmer water to the molecules in the cooler water until they have the same average energy (temperature). Since the two samples of water are identical in volume, the thermal equilibrium that is reached is an average of the two temperatures.

Curricular and Instructional Considerations

Elementary Students

At the elementary level, students' experiences with materials are primarily observational. They know how to measure temperature with a thermometer. Mixing hot and cold water and predicting and observing the resulting temperature is observational and should initially be approached qualitatively. At this level, students should not be expected to explain the resulting temperatures in terms of energy transfer. The emphasis should be on exploring how heat spreads from one place to another, including how cooler materials can become warmer and vice versa. The basic concept of conduction can be introduced at the level of objects or materials, but not atoms or molecules. Although students at this grade level are not ready to understand what heat is or the difference between heat and temperature, it is helpful to refer to energy transfers in terms of gaining or losing heat in order to help students overcome their intuitive notion that cold is a substance that spreads like heat.

Middle School Students

In middle school, students shift their focus from observing what happens when warm and cold water are mixed together to explaining what happens in terms of the flow of heat from warm to cold. This is also a time when the term *thermal energy,* as opposed to *heat energy,* is introduced. Students begin to connect the idea of heat with a movement of energy. As upper middle school students develop a model of particle motion as it relates to heat and temperature, they can begin to connect this idea to the concept of conduction as a mechanism for the transfer of heat between atoms or molecules. The idea of energy transfer begins to become quantitative as students see that an energy loss in one material is a gain in energy for the other, and this idea leads to the development of an understanding of the key principle of conservation of energy. This probe targets the grade-level expectation of understanding that heat flows from warmer to cooler objects and materials until they both reach the same temperature.

High School Students

Students build on their experiences with energy transfers in middle school to investigate a variety of energy transfers more systematically and quantitatively, collecting evidence that confirms that energy is conserved during energy transfers and recognizing the loss of some thermal energy through dissipation. With careful instruction, students can begin to understand the operational distinction between heat and temperature, but the idea of heat as random motion and vibrating atoms or molecules is still difficult for many students to grasp.

Administering the Probe

You may wish to use visual props for this probe to demonstrate the two equal volumes, the pouring of one glass into another, and mixing the combination of the two samples. With younger students, you can change the probe to be more qualitative by asking what would happen if cool and warm water were mixed together—with the choices being (1) the water will become even cooler, (2) the water will become even warmer, or (3) the water will be somewhere in between the two. The probe can be further adapted for older students by changing the volumes of water that are mixed together.

Related Ideas in *National Science Education Standards* (NRC 1996)

. . . .

K–4 Light, Heat, Electricity, and Magnetism

★ Heat can move from one object to another by conduction.

5–8 Transfer of Energy

• Energy is transferred in many ways.
★ Heat moves in predictable ways, flowing from warmer objects to cooler ones until both reach the same temperature.

9–12 Conservation of Energy and the Increase in Disorder

• Heat consists of random motion and the vibrations of atoms, molecules, and ions. The higher the temperature, the greater the atomic or molecular motion.

• Everything tends to become less organized and less orderly over time. Thus in all energy transfers, the overall effect is that the energy is spread out uniformly. Examples are the transfer of energy from hotter to cooler objects by conduction, radiation, or convection and the warming of our surroundings when we burn fuels.

Related Ideas in *Benchmarks for Science Literacy* (AAAS 1993)

. . . .

3–5 Energy Transformations

★ When warmer things are put with cooler ones, the warm ones lose heat and the cool ones gain it until they are all at the same temperature. A warmer object can warm a cooler one by contact or at a distance.

6–8 Structure of Matter

• Atoms and molecules are perpetually in motion. Increased temperature means greater average energy of motion.

6–8 Energy Transformations

★ Thermal energy is transferred through a material by the collisions of atoms within the material.
• Energy appears in different forms and can be transformed within a system.

9–12 Energy Transformations

• Whenever the amount of energy in one place or form diminishes, the amount

★ Indicates a strong match between the ideas elicited by the probe and a national standard's learning goal.

in other places or forms increases by the same amount.

★ Thermal energy in a system is associated with the disordered motions of its atoms or molecules. In any system of atoms or molecules, the statistical odds are that the atoms or molecules will end up with less order than they originally had and that the thermal energy will be spread out more evenly.

Related Research

- Middle school students often do not explain the process of heating and cooling in terms of heat being transferred. When transfer ideas are involved, some students will think that cold is being transferred from a colder to warmer object. Other students think that both heat and cold are transferred at the same time (AAAS 1993).

- Middle and high school students do not always explain heat-exchange phenomena as interactions. For example, students may describe objects as tending to cool down or release heat spontaneously without acknowledging that the object has come in contact with a cooler object or area (AAAS 1993).

- Numerous studies have shown that few middle and high school students understand the molecular basis of heat transfer after instruction. Difficulties in understanding remain even with instruction that is specially designed to explicitly address the difficulty of understanding heat

transfer (AAAS 1993).

- Researchers have found that difficulties experienced by students in response to questions that ask them to predict the final temperature of a mixture of two quantities of water, given the initial temperature of the components, depend on the form in which the temperature problems are presented. Qualitative tasks in which the water is described as warm, cool, hot, or cold are easier than quantitative ones in which specific temperatures are given. The mixing of waters at different temperatures (e.g., hot and cold or 30°C and 80°C) is more difficult than mixing water at the same temperature (e.g., warm and warm or 50°C and 50°C). It is not until around the age of 12 that most students can predict quantitatively what will happen in the type of problem posed in this probe (Erickson and Tiberghien 1985).

- Student responses to tasks similar to the one posed in this probe have been categorized according to the strategy used. Younger students (ages 8–9) prefer an addition strategy, whereas older students are more apt to use a subtraction strategy, which at least acknowledges that the final temperature lies somewhere in between. However, students ages 12–16 were as likely to use an addition or subtraction strategy as to use an averaging strategy (Erickson and Tiberghien 1985).

- When considering the final temperature of two beakers of cold water at the same temperature mixed together, children ages 4–6 often judge the temperature to be the same. However, children ages 5–8 often say that the water will be twice as cold because there is twice as much water. At age 12, students describe the water as being the same temperature when mixed together, much like the very young children. One possible explanation for this progression is that young children do not consider amount and judge temperature as if it were an extensive physical quantity. Older children are more able to differentiate between intensive and extensive quantities, understanding that temperature remains unchanged despite the amount of water. It was also found that children tended to make more correct predictions of temperature when equal amounts of hot and cold water were mixed than when two equal amounts of cold water were mixed (Driver et al. 1994).

Suggestions for Instruction and Assessment

- This probe can be followed up with an inquiry-based investigation. Ask the question, encourage students to commit to a prediction, then test it with the temperatures stated in the probe (use caution when students are handling hot liquids). The dissonance involved in discovering that their predictions and results may differ should lead to testing other combinations of temperatures, including mixing of water at the same temperature.

- Depending on the age of the students, vary their experiences to include mixing same temperatures; mixing samples at two different cold, hot, warm, or cool temperatures; mixing two different temperatures that vary by less than 10 degrees or more than 50 degrees; mixing unequal volumes at same temperatures and unequal volumes at different temperatures; mixing three of four different samples at same and different volumes; and so on. Ideally, have students come up with the various configurations to test. Have students discover the pattern that results from a variety of mixings and use their discovery to lead into an explanation of heat transfer.

- Try juxtaposing two different representational systems. Give one probe in which the prediction is stated as mixing equal amounts of cold and hot water and give the other stated in quantitative terms as in this probe. If their responses differ, use this conflict-inducing strategy to begin to help students distinguish between the ideas of temperature and heat.

- To develop the idea of conduction, provide students with multiple opportunities to mix hot and cold objects and materials, not only liquids. For example, freeze or heat solid objects and add them to a container of water. Do not end activities with findings; be sure to engage students in discussion about their findings and explanations, connecting them to the variety of experiences they have had in mixing objects and materials at different temperatures in order

11

to develop generalizations and appreciate their range of applications.

Related NSTA Science Store Publications and NSTA Journal Articles

American Association for the Advancement of Science (AAAS). 1993. *Benchmarks for science literacy.* New York: Oxford University Press.

Driver, R., A. Squires, P. Rushworth, and V. Wood-Robinson. 1994. *Making sense of secondary science: Research into children's ideas.* London: RoutledgeFalmer.

Keeley, P. 2005. *Science curriculum topic study: Bridging the gap between standards and practice.* Thousand Oaks, CA: Corwin Press.

National Research Council (NRC). 1996. *National science education standards.* Washington, DC: National Academy Press.

Robertson, W. 2002. *Energy: Stop faking it! Finally understanding science so you can teach it.* Arlington, VA: NSTA Press.

Welborn, J. 2002. Feel the heat. *Science Scope* (Nov./Dec.): 51–52.

References

American Association for the Advancement of Science (AAAS). 1993. *Benchmarks for science literacy.* New York: Oxford University Press.

Driver, R., A. Squires, P. Rushworth, and V. Wood-Robinson. 1994. *Making sense of secondary science: Research into children's ideas.* London: RoutledgeFalmer.

Erickson, G., and A. Tiberghien, A. 1985. Heat and temperature. In *Children's ideas in science,* eds. R. Driver, E. Guesne, and A. Tiberghien, 52–84. Milton Keynes, England: Open University Press.

Keeley, P. 2005. *Science curriculum topic study: Bridging the gap between standards and practice.* Thousand Oaks, CA: Corwin Press.

National Research Council (NRC). 1996. *National science education standards.* Washington, DC: National Academy Press.

Related Curriculum Topic Study Guides
(Keeley 2005)
"Energy"
"Heat and Temperature"
"Physical Properties and Change"

Life Science Assessment Probes

Probes

Core Science Concepts	Plants					Genetics	Cells	Adaptation
	Is It a Plant?	Need of Seeds	Plants in the Dark and Light	Is It Food for Plants?	Giant Sequoia Tree	Baby Mice	Whale and Shrew	Habitat Change
Adaption								✓
Behavioral response			✓					✓
Biological Classification	✓							✓
Cell Division							✓	
Cells							✓	
Chromosomes						✓		
Food				✓				
Genes						✓		
Germination		✓						
Growth			✓				✓	
Inherited Traits						✓		
Life Cycles		✓						
Needs of Organisms		✓	✓					
Photosynthesis				✓	✓			
Plants	✓		✓	✓	✓			
Seeds		✓						
Transformation of Matter					✓			

Is It a Plant?

Each of the things listed below can be found living and growing in its environment. Put an X next to the things that you consider to be plants.

___ fern   ___ grass   ___ moss

___ vine   ___ grasshopper   ___ tomato

___ mold   ___ flower   ___ tree

___ onion   ___ weed   ___ bush

___ cactus   ___ bacteria   ___ mushroom

___ cabbage   ___ dandelion   ___ carrot

Explain your thinking. Describe the "rule" or reasoning you used to decide if something is a plant.

_____

_____

_____

_____

_____

_____

_____

_____

_____

_____

Is It a Plant?

Teacher Notes

Purpose

The purpose of this assessment probe is to elicit students' ideas about plants. The probe is designed to determine how students classify organisms as plants.

Related Concepts

biological classification, plants

Explanation

The items on the list biologically considered to be plants are fern, grass, moss, vine, tomato, flower, tree, onion, weed, bush, cactus, cabbage, dandelion, and carrot. Plants are multicellular organisms that are generally able to make their own food through photosynthesis. Many plant cells contain pigments capable of absorbing light. Plants have a structure called a cell wall, which is made mostly of cellulose. Plants can vary in size, from tall trees to short mosses.

Some of the items on this list are "plant-like" but are not plants. For example, mushrooms and mold are classified as fungi. Their cell walls are generally made of chitin instead of cellulose, and they do not make their own food or contain light-absorbing pigments within their cells. The grasshopper is a green animal; some students think that anything green is a plant. Bacteria are prokaryotic organisms, meaning they lack membrane-bound organelles like those found in plants, even though bacteria have a cell wall and some bacteria are capable of photosynthesis.

Curricular and Instructional Considerations

Elementary Students

Plants are a common topic for investigation into the characteristics of and processes that maintain life. Typically, young students learn to distinguish plants from other organisms by their structures, unique needs (such as light), observable functions, and outward appearance (green). Characteristics used for grouping plants or distinguishing plants from other organisms are based primarily on observations of their external structures and characteristics. Details about their cell structures, photosynthesis, embryological development, and modern taxonomy exceed what is developmentally appropriate at this level.

Although students have many opportunities to observe and investigate plants and plant parts, their conception of a plant may be limited to the types of plants they have experiences with—typically, flowering plants. Students may fail to develop a generalization of what a plant is if they are limited in experience to one type of plant. This probe is useful in determining whether students recognize that plants are a broad category for a variety of biologically diverse organisms. Elementary students should not be expected to define a plant by sophisticated structural, functional, and developmental criteria.

Middle School Students

Middle school is the time when students begin to use scientific classification criteria to distinguish among the major kingdoms or domains of organisms. They distinguish plants from other plant-like organisms, including fungi and green algae. They begin to learn about bacteria as simple one-celled organisms that differ from plants, fungi, protists, and animals. They examine different types of cells with simple microscopes and can distinguish plant cells from animal cells by their observable structures. They are familiar with plant cell structures such as cell walls and chloroplasts but do not need to know the structural components. Students recognize the ability of most plants to make their own food through photosynthesis.

This probe is useful in determining whether students have a generalized concept of plants and in examining whether they use more sophisticated biological criteria than elementary students in distinguishing plants from other organisms.

High School Students

At the high school level, students use modern taxonomies to distinguish among and between organisms. They are introduced to the three major domains of modern taxonomy and the kingdoms, such as Plantae, within those domains. At this level they can begin to use more sophisticated criteria to define plants, including embryological development, plastids, autotrophism, and molecular substances found in plant structures. However, caution must be used with the extensive terminology students typically encounter in biology. Even though they may "learn" these terms and criteria, they may still revert to their earlier conceptions of what a plant is.

Administering the Probe

Be sure students understand that the terms on the list refer to the complete organism. For example, *tomato* refers to the complete tomato plant, not just the fruit. Make sure students are familiar with the items on the list; you may wish to remove or replace items that students have little or no familiarity with.

This probe can be used as a card sort. In small groups, students can sort cards with the names of organisms into two groups—"plants" and "not plants." Listening carefully to students as they decide or argue about which category the organism on the list belongs to lends additional insight into student thinking. An alternative sorting method is to use pictures of plants with or without names, including plants in natural settings, gardens, and pots. Pictures can also be combined with real examples of plants on the list. With early elementary students, sorting can be done as a whole-class discussion. High school teachers may want to add more sophisticated examples that students have previously encountered, such as microscopic algae, giant kelp, yeast, euglena, and Venus flytrap.

Related Ideas in *National Science Education Standards* (NRC 1996)

. . . .
K–4 The Characteristics of Organisms

- Organisms have basic needs. For example, animals need air, water, and food; plants require air, water, nutrients, and light.
- Each plant or animal has different structures that serve different functions in growth, survival, and reproduction.

5–8 Diversity and Adaptations of Organisms

★ Millions of species of animals, plants, and microorganisms are alive today. Although different species might look dissimilar, the unity among organisms becomes apparent from an analysis of internal structures, the similarity of their chemical processes, and the evidence of common ancestry.

5–8 Populations and Ecosystems

★ Populations of organisms can be categorized by the function they serve in an ecosystem. Plants and some microorganisms are producers—they make their own food. All animals, including humans, are consumers, which obtain food by eating other organisms. Decomposers, primarily bacteria and fungi, are consumers that use waste materials and dead organisms for food.

9–12 The Cell

★ Plant cells contain chloroplasts, the site of photosynthesis. Plants and many microorganisms use solar energy to combine molecules of carbon dioxide and water into complex, energy-rich organic compounds and release oxygen to the environment.

★ Indicates a strong match between the ideas elicited by the probe and a national standard's learning goal.

9–12 Biological Evolution

☆ Biological classifications are based on how organisms are related. Organisms are classified into a hierarchy of groups and subgroups based on similarities that reflect their evolutionary relationships.

Related Ideas in *Benchmarks for Science Literacy* (AAAS 1993)

K–2 Diversity of Life

• Some animals and plants are alike in the way they look and in the things they do, and others are very different from one another.

3–5 Diversity of Life

☆ A great variety of living things can be sorted into groups in many ways using various features to decide which things belong to which group.

6–8 Diversity of Life

☆ One of the most general distinctions among organisms is between plants, which use sunlight to make their own food, and animals, which consume energy-rich foods. Some kinds of organisms, many of them microscopic, cannot be neatly classified as either plants or animals.

☆ Similarities among organisms are found in internal anatomical features, which can be used to infer the degree of relatedness among organisms. In classifying organisms, biologists consider details of internal

and external structures to be more important than behavior or general appearance.

Related Research

• Some research indicates that in second grade students begin to shift their thinking about organisms based on perceptual and behavioral features to more biological representations (AAAS 1993).

• Methods of grouping organisms vary by developmental level. For example, in upper elementary school some students may group organisms, such as plants, by observable features, whereas others base their groupings on concepts. By middle school, students start to group organisms hierarchically when they are asked to do so. It is not until high school that students use hierarchical taxonomies without being prompted (AAAS 1993).

• Elementary and middle school students hold a more restricted meaning for the word *plant* than biologists. Trees, vegetables, and grass are often not considered to be plants (AAAS 1993).

• In a study by Leach et al. (1992), students used "plant," "tree," and "flower" as mutually exclusive groups. However, when students were given a restricted number of classification categories in a classification task, they assigned trees and flowers to the plant category (Driver et al. 1994).

• Ryman (1974) found that 12-year-old English students had more difficulty classifying plants into taxonomic categories than they did classifying animals. It appeared that students learned a "school science" way of classifying but retained their intuitive ideas about plant classification in everyday life (Driver et al. 1994).

- In a study by Stead (1980), some children suggested that a plant is something that is cultivated; hence grass and dandelions were considered weeds, not plants. Some children considered cabbage and carrots to be vegetables, not plants. They viewed vegetables as a comparable set rather than a subset of plants.

Suggestions for Instruction and Assessment

- Provide opportunities for students to observe, identify, and investigate a variety of flowering and nonflowering plants, not just the typical flowering plants (e.g., bean plants) that are commonly used in classroom investigations. Students should investigate vegetables, flowers, ferns, mosses, trees, vines, weeds, bushes, and grasses.

- Alert students to the common use of the word *plant* versus the scientific meaning of the word, so that students will recognize trees, weeds, vines, and other plants that are referred to by those names as also belonging to a larger group called plants.

- At the high school level, help students reflect back on their growing knowledge of plant taxonomy from the macroscopic level (organism) to the microscopic level (plant cell and cell parts) to the molecular level (chlorophyll).

- Have students develop pre-instruction and post-instruction concept maps to illustrate their meaning of the concept of a plant.

- Draw comparisons between familiar subsets of animal classifications (e.g., reptiles, amphibians, fish, birds, and mammals grouped under vertebrates, which are grouped under animals) and plant classifications so that students can develop the idea of "plant" as a broad category that includes a variety of subsets with common characteristics. For example, there are vascular plants and nonvascular plants. Vascular plants can be broken down into subsets that include plants that do not produce seeds and those that produce seeds. Seed-producing plants can be further broken down into subsets of flowering seed plants and nonflowering seed plants. This will help older students understand that plants include a variety of groupings.

Related NSTA Science Store Publications and Journal Articles

American Association for the Advancement of Science (AAAS). 1993. *Benchmarks for science literacy.* New York: Oxford University Press.

Barman, C., M. Stein, N. Barman, and S. McNair. 2003. Students' ideas about plants: Results from a national study. *Science & Children* 41 (1): 46–51.

Driver, R., A. Squires, P. Rushworth, and V. Wood-Robinson. 1994. *Making sense of secondary science: Research into children's ideas.* London: RoutledgeFalmer.

Franklin, K. 2001. Bring classification to life. *Sci-*

ence Scope 25 (3): 36–41.

Keeley, P. 2005. *Science curriculum topic study: Bridging the gap between standards and practice.* Thousand Oaks, CA: Corwin Press.

Lawniczak, S., D. T. Gerber, and J. Beck. 1994. Plants on display. *Science & Children* 41 (9): 24–29.

National Research Council (NRC). 1996. *National science education standards.* Washington, DC: National Academy Press.

Texley, J. 2002. Teaching the new taxonomy: Getting up to speed on recent developments in taxonomy. *The Science Teacher* 69 (3): 62–66.

Related Curriculum Topic Study Guides

(Keeley 2005)

"Biological Classification"

"Plant Life"

References

American Association for the Advancement of Science (AAAS). 1993. *Benchmarks for science literacy.* New York: Oxford University Press.

Driver, R., A. Squires, P. Rushworth, and V. Wood-Robinson. 1994. *Making sense of secondary science: Research into children's ideas.* London: RoutledgeFalmer.

Keeley, P. 2005. *Science curriculum topic study: Bridging the gap between standards and practice.* Thousand Oaks, CA: Corwin Press.

Leach, J., R. Driver, P. Scott, and C. Wood-Robinson. 1992. *Progression in conceptual understanding of ecological concepts by pupils aged 5-16.* Leeds, England: University of Leeds, Centre for Studies in Science and Mathematics Education.

National Research Council (NRC). 1996. *National science education standards.* Washington, DC: National Academy Press.

Ryman, D. 1974. Children's understanding of the classification of living organisms. *Journal of Biological Education* 23 (3): 199–207.

Stead, B. 1980. *Plants.* LISP Working Paper 24. Hamilton, New Zealand: University of Waikato, Science Education Research Unit.

13

Needs of Seeds

Seeds sprout and eventually grow into young plants called seedlings. Put an X next to the things you think a seed needs in order for it to sprout.

___ water

___ soil

___ air

___ food

___ sunlight

___ darkness

___ warmth

___ Earth's gravity

___ fertilizer

Explain your thinking. Describe the "rule" or reasoning you used to decide what a seed needs in order to sprout.

_____

_____

_____

_____

_____

_____

_____

_____

Needs of Seeds

Teacher Notes

Purpose

The purpose of this assessment probe is to elicit students' ideas about seeds. It specifically probes to find out if students recognize that a seed has needs, similar to other organisms, that allow it to develop into the next stage of its life cycle.

Related Concepts

germination, life cycles, needs of organisms, seeds

Explanation

The best response is that seeds need water, air, and warmth. Like all living things, the plant embryo inside a seed needs water, air, food, and warmth to carry out the life processes that will support its germination and growth. The young plant embryo needs food as its source of energy and building material for growth. However, the food it needs is already contained within the seed in the form of a cotyledon, since

a young sprout does not yet have the leaves to enable it to carry out photosynthesis. Air is necessary for seeds to respire. Seeds must take in oxygen to use and release energy from their food. Seeds also require a warm temperature and water for the life-sustaining chemical reactions that take place in the cells of the young plant embryo to occur. However, some seeds, such as acorns, need to go through a cold period before they germinate. Too much liquid water "drowns" seeds by preventing them from taking in oxygen and causes them to rot. Some seeds can sprout in humid air without the need for a moist surface. Hence, the right amount of water needs to be available.

Seeds can sprout without soil as long as they have a source of moisture. Sunlight is not needed, as evidenced by the way many seeds germinate when covered by soil. Seeds have sprouted in microgravity in space. Gravity affects the ability of the sprout to send its early root structures downward, but seeds can sprout even in conditions where gravity is much less than that on Earth. Fertilizers are not needed by seeds. They are used by plants once they have established roots and can take in these substances from the soil to contribute essential elements to the cells that make up their plant structures.

Curricular and Instructional Considerations

Elementary Students

Elementary students typically have experiences germinating seeds and growing plants. Early experiences focus primarily on the seed's need for water and warmth. Since students often plant their seeds in soil and water them, they may not realize that soil is not necessary for a seed to germinate. Likewise, since the seeds are in soil, they may think darkness is a requirement and that sunlight would harm a seed. Investigations that involve germinating seeds under various conditions help students recognize that some factors are needed for germination and others are not. Students can eventually distinguish between the needs of seeds and the needs of the growing plant.

Middle School Students

Middle school students typically have more systematic experiences investigating plants and their needs. The seed's cotyledon is recognized and investigated as a source of food

for the developing sprout and seedling before it grows into a plant capable of making food from carbon dioxide and water using energy from sunlight. As students develop an understanding that all living things carry out similar life processes, they recognize that seeds also need oxygen as well to carry out cellular respiration. At this level students should be able to distinguish between what seeds need to germinate and what complete plants need to function.

High School Students

Although basic germination experiences take place in elementary and middle school, this probe may be useful in determining if students retain misconceptions related to germination. High school students learn about specialized factors that can affect germination, such as the need for some seeds to travel through animals' digestive systems in order to open the seed coat or the need for some conifers to be exposed to fire in order to release seeds. They may also investigate the concept of inhibitors where chemicals released by some plants will inhibit the germination of other seeds in their area.

Administering the Probe

You may wish to use visual props with the probe. Show students an ungerminated bean seed and a germinated seed or show them a picture of a sprout or an actual sprout if there are students who do not know what a sprout is. For older students you may substitute the word

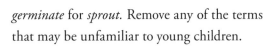

germinate for *sprout*. Remove any of the terms that may be unfamiliar to young children.

This probe can also be used as a card sort. Write the words on cards and have students sort them into piles of things seeds need to sprout and things seeds do not need to sprout. Listen carefully as they discuss their ideas about which pile to put their cards in.

Related Ideas in *National Science Education Standards* (NRC 1996)

. . . .

K–4 The Characteristics of Organisms

★ Organisms have basic needs. For example, plants require air, water, nutrients, and light.

K–4 Life Cycles of Organisms

• Plants and animals have life cycles that include being born, developing into adults, reproducing, and eventually dying. The details of this life cycle are different for different organisms.

6–8 Regulation and Behavior

• All organisms must be able to obtain and use resources, grow, reproduce, and maintain stable internal conditions while living in a constantly changing external environment.

Related Ideas in *Benchmarks for Science Literacy* (AAAS 1993)

. . . .

K–2 Cells

★ Most living things need water, food, and air.

K–2 Flow of Matter and Energy

★ Plants and animals both need to take in water, and animals need to take in food. In addition, plants need light.

3–5 Flow of Matter and Energy

• Some source of "energy" is needed for all organisms to stay alive and grow.

6–8 Flow of Matter and Energy

• Food provides the fuel and building material for all organisms. Plants use the energy from light to make sugars from carbon dioxide and water. This food can be used immediately or stored for later use.

Related Research

• Many children think that plants always need light to grow, and they apply this idea to germination (Driver et al. 1994).

• Driver et al.'s study of a large sample of 15-year-olds showed that many of the students thought that respiration only occurred in the cells of leaves of plants since these cells have gas exchange pores. They did not see things like seeds as exchanging gases (Driver et al. 1994).

• Some students fail to recognize a seed as a living thing; therefore they do not recognize that seeds have needs similar to those of other living things (Driver et al. 1994).

• Students appear to believe that food and light are necessary for all stages of plant

---

★ Indicates a strong match between the ideas elicited by the probe and a national standard's learning goal.

growth. However, prior to instruction they often do not understand that light is a requirement for food making but not a requirement for growth. A study conducted by Roth, Smith, and Anderson (1983) found that students held strongly to the idea that light is always required by plants even in the face of contrary evidence such as seedlings germinating in the dark (Driver et al. 1994).

- Russell and Watt (1990) interviewed younger students about their ideas related to conditions for growth, focusing on germinations as well as vegetative growth. Ninety percent of the 60 children interviewed identified water as necessary. Only a few mentioned air, gases, "food" (which to them was soil nutrients), sun, light, or heat (Driver et al. 1994).

- Some students have difficulty distinguishing between germination and vegetative growth (Driver et al. 1994).

Suggestions for Instruction and Assessment

- This probe could be followed up with an inquiry-based investigation. Have students make predictions and test their ideas with seeds that germinate easily, such as bean seeds.

- Use caution when teaching the ideas in the K–4 standards that state that all plants need light. This is true for part of their life cycle, but a plant embryo, a sprout, and an emerging seedling do not need light at those stages in the life cycle because they have a stored source of energy. Once it has

used up all the food that was stored in the seed's cotyledon, the seedling needs light to make its own food, using its true leaves.

- Examine seeds, helping students see where water is taken in, gases are exchanged, and food is stored for the young embryo. Rather than focusing on naming the parts of a seed, help students understand how the seed contributes to the growth and life functions of the young plant.

Related NSTA Science Store Publications and Journal Articles

American Association for the Advancement of Science (AAAS). 1993. *Benchmarks for science literacy.* New York: Oxford University Press.

Barman, C., M. Stein, N. Barman, and S. McNair. 2003. Students' ideas about plants: Results from a national study. *Science & Children* 41 (1): 46–51.

Cavallo, A. 2005. Cycling through plants. *Science & Children* (Apr./May): 22–27. Also available online in *NSTA WebNews Digest* at *www.nsta. org/main/news/stories/science_story.php?news_ story_ID=50416.*

Driver, R., A. Squires, P. Rushworth, and V. Wood-Robinson. 1994. *Making sense of secondary science: Research into children's ideas.* London: RoutledgeFalmer.

Keeley, P. 2005. *Science curriculum topic study: Bridging the gap between standards and practice.* Thousand Oaks, CA: Corwin Press.

National Research Council (NRC). 1996. *National science education standards.* Washington, DC: National Academy Press.

Quinones, C., and B. Jeanpierre. 2005. Planting the spirit of inquiry. *Science & Children* 42 (7): 32–35.

West, D. 2004. Bean plants: A growth experience. *Science Scope* (Apr.): 44–47.

Related Curriculum Topic Study Guides

(Keeley 2005)

"Life Processes and Needs of Organisms"

"Plant Life"

"Reproduction, Growth, and Development (Life Cycles)"

References

American Association for the Advancement of Science (AAAS). 1993. *Benchmarks for science literacy.* New York: Oxford University Press.

Driver, R., A. Squires, P. Rushworth, and V. Wood-Robinson. 1994. *Making sense of secondary science: Research into children's ideas.* London: RoutledgeFalmer.

Keeley, P. 2005. *Science curriculum topic study: Bridging the gap between standards and practice.* Thousand Oaks, CA: Corwin Press.

National Research Council (NRC). 1996. *National science education standards.* Washington, DC: National Academy Press.

Roth, K., E. Smith, and C. Anderson. 1983. *Students' conceptions of photosynthesis and food for plants.* East Lansing, MI: Michigan State University, Institute for Research on Teaching.

Russell, T., and D. Watt. 1990. *SPACE research report: Growth.* Liverpool, England: Liverpool University Press.

Plants in the Dark and Light

Four friends wondered how light affected the growth of plants. They decided to test their ideas using young bean plants. One set of plants was put in a dark closet for eight days. The other set of plants was put on a shelf near a sunny window for eight days. The friends then measured the height of the plants after eight days. This is what they predicted:

Carl: "I think the plants in the dark closet will be the tallest."

Monique: "I think the plants by the sunny window will be the tallest."

Jasmine: "I think the plants will be about the same height."

Drew: "I think the plants in the closet will stop growing and die."

Which friend do you agree with and why? Explain your thinking.

_____

_____

_____

_____

_____

_____

_____

Plants in the Dark and Light

Teacher Notes

Purpose

The purpose of this assessment probe is to elicit students' ideas about plant growth. It specifically probes to find out if students think plants only grow if they are exposed to light.

Related Concepts

behavioral response, growth, needs of organisms, plants

Explanation

There is no one correct answer for this probe because plant growth depends on several conditions. Plants may grow taller in a dark place for a while: They respond to the lack of light by growing "taller" and more spindly, and the

plant stem and leaves may be yellow and not as leafy. The growth in the dark is caused by *auxins,* which are substances that regulate plant growth. Auxins are found in young tissue called the apical meristem, at the end of a shoot or stem, and are transported downward from the tip of the stem or shoot. Auxins stimulate plant cells to elongate, resulting in an increase in plant height.

Light is the form of energy plants use to make food from carbon dioxide and water. This food can be used to carry out life processes and to build new structures for growth or repair, or it can be stored for later use. The plant in the dark uses the food it has made to continue growing taller as the cells continue

to elongate. It may grow faster than the plant on the windowsill, although the growth will be spindly and etiolated. If the plant is in the dark for an extended period of time, eventually its food will be used up and the plant will no longer have the food energy and building material it needs to live and grow.

Curricular and Instructional Considerations

Elementary Students

Elementary students have varied experiences investigating the growth of plants. Knowing that plants require light is a grade-level expectation in the national standards. However, it is not until middle school that students begin to understand why plants need light beyond knowing that they need it to survive. The idea that light is needed for survival may imply that plants stop growing and soon die when placed in the dark. This probe is useful in determining whether students think that the short-term absence of light will result in the immediate death or lack of growth of the plant.

Middle School Students

In middle school, students design their own experiments with plants that allow students to identify, manipulate, and control variables. They develop an understanding that plants make food and that this food can be stored and used by a plant when needed. They also begin to learn about behavioral responses of animals and plants, including plant tropisms. This probe is useful in determining whether students can link ideas about stored food to what happens when plants are deprived of light for a short period of time.

High School Students

At this level, students are more systematic in investigating plant functions. Their knowledge of plant physiology and behavioral response includes the role of auxins and plant tropisms. Even though students may understand how plants respond to a lack of sunlight, they may still revert to their intuitive beliefs about plants being unable to grow in the dark.

Administering the Probe

Make sure students can visualize the plants on the windowsill and the plants in the closet. You might use a prop to demonstrate the set-up. You might ask students to draw a picture of what a plant would look like in each situation before and after the experiment.

Related Ideas in *National Science Education Standards* (NRC 1996)

K–4 The Characteristics of Organisms

- Organisms have basic needs. For example, plants require air, water, nutrients, and light.
- The behavior of individual organisms is influenced by internal cues (such as hunger) and by external cues (such as a change in the environment).

6–8 Regulation and Behavior

★ All organisms must be able to obtain and use

★ Indicates a strong match between the ideas elicited by the probe and a national standard's learning goal.

resources, grow, reproduce, and maintain stable internal conditions while living in a constantly changing external environment.

★ Behavior is one kind of response an organism can make to a internal or environmental stimulus.

9–12 The Behavior of Organisms

★ Organisms have behavioral responses to internal changes and to external stimuli. Responses to external stimuli can result from interactions with the organism's own species and others, as well as environmental changes; these responses can either be innate or learned. The broad patterns of behavior exhibited by animals have evolved to ensure reproductive success. Animals often live in unpredictable environments, and so their behavior must be flexible enough to deal with uncertainty and change. Plants also respond to stimuli.

Related Ideas in *Benchmarks for Science Literacy* (AAAS 1993)

. . . .

K–2 Flow of Matter and Energy

• Plants and animals both need to take in water, and animals need to take in food. In addition, plants need light.

Related Research

• Students appear to accept the idea that light is needed for all the stages of plant growth. However, they may not understand that light is used to make food for

the plant and is not a condition for growth itself. A study conducted by Roth, Smith, and Anderson (1983) found that students held strongly to the idea that light is always required by plants even in the face of contrary evidence such as plants growing taller in the dark (Driver et al. 1994).

• In a study conducted by Wandersee (1983), secondary students were asked to draw their predictions for a plant that was grown in a dark cupboard and one that was kept on a windowsill where the light could shine through. Almost 90% of the students drew the plant on the windowsill as large and healthy, leaning toward the light—which showed some understanding of phototropism. Eighty-five percent of the students drew the plant in the cupboard as being stunted. Only 11% of the students drew the plant in the cupboard as tall and spindly (Driver et al. 1994).

Suggestions for Instruction and Assessment

• This probe can be followed up with an inquiry-based investigation in which students test their prediction. Once they have gathered and analyzed their findings, encourage them to explain their findings. Challenge them to revise their initial ideas based on their evidence.

• Provide students with an opportunity to test their ideas with different types of plants so that their ideas about plant growth are not limited to a particular type of plant.

• Ask students to describe situations where

★ Indicates a strong match between the ideas elicited by the probe and a national standard's learning goal.

they have seen plants growing in the absence of light, such as grass under a board or a houseplant left in a basement. Describe how the plant looks compared with a plant grown under ordinary light conditions.

- Be explicit when developing the idea that plants need light to make food. Clarify the need for light to make food versus the more commonly held idea of light as a general need at all times.

Related NSTA Science Store Publications and Journal Articles

American Association for the Advancement of Science (AAAS). 1993. *Benchmarks for science literacy.* New York: Oxford University Press.

Barman, C., M. Stein, N. Barman, and S. McNair. 2003. Students' ideas about plants: Results from a national study. *Science & Children* 41 (1): 46–51.

Cavallo, A. 2005. Cycling through plants. *Science & Children* (Apr./.May): 22–27.

Driver, R., A. Squires, P. Rushworth, and V. Wood-Robinson. 1994. *Making sense of secondary science: Research into children's ideas.* London: RoutledgeFalmer.

Keeley, P. 2005. *Science curriculum topic study: Bridging the gap between standards and practice.* Thousand Oaks, CA: Corwin Press.

National Research Council (NRC). 1996. *National science education standards.* Washington, DC: National Academy Press.

Quinones, C., and B. Jeanpierre. 2005. Planting the spirit of inquiry. *Science & Children* 42 (7): 32–35.

West, D. 2004. Bean plants: A growth experience. *Science Scope* (Apr.): 44–47.

Related Curriculum Topic Study Guides

(Keeley 2005)
"Photosynthesis and Respiration"
"Plant Life"
"Regulation and Control"

References

American Association for the Advancement of Science (AAAS). 1993. *Benchmarks for science literacy.* New York: Oxford University Press.

Driver, R., A. Squires, P. Rushworth, and V. Wood-Robinson. 1994. *Making sense of secondary science: Research into children's ideas.* London: RoutledgeFalmer.

Keeley, P. 2005. *Science curriculum topic study: Bridging the gap between standards and practice.* Thousand Oaks, CA: Corwin Press.

National Research Council (NRC). 1996. *National science education standards.* Washington, DC: National Academy Press.

Roth, K., E. Smith, and C. Anderson. 1983. *Students' conceptions of photosynthesis and food for plants.* East Lansing, MI: Michigan State University, Institute for Research on Teaching.

Wandersee, J. 1983. Students' misconceptions about photosynthesis: A cross-age study. In *Proceedings of the international seminar: Misconceptions in science and mathematics,* eds. H. Helm and J. Novak, 441–446. Ithaca, NY: Cornell University.

Is It Food for Plants?

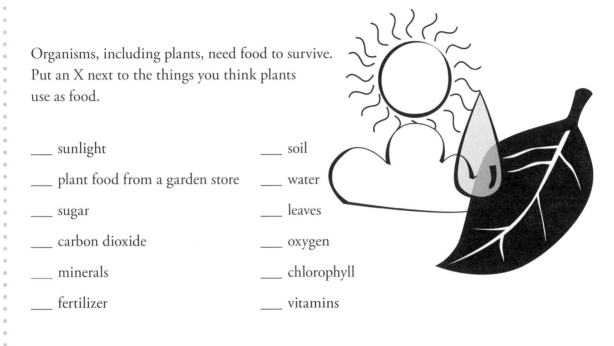

Organisms, including plants, need food to survive. Put an X next to the things you think plants use as food.

___ sunlight   ___ soil

___ plant food from a garden store   ___ water

___ sugar   ___ leaves

___ carbon dioxide   ___ oxygen

___ minerals   ___ chlorophyll

___ fertilizer   ___ vitamins

Explain your thinking. How did you decide if something on the list is food for plants?

_____

_____

_____

_____

_____

_____

_____

_____

_____

_____

Is It Food for Plants?

Teacher Notes

Purpose

The purpose of this assessment probe is to elicit students' ideas about food and plants. The probe is designed to reveal whether students use a biological concept of food to identify what plants use for food.

Related Concepts

food, photosynthesis, plants

Explanation

The best response is sugar. Sugars, such as glucose, are simple carbohydrates made and used by plants as food. Plants differ from animals in that they are able to use the energy from sunlight to transform inorganic carbon dioxide and water, which they take in from their environment, into food. This process is called photosynthesis. Part of the confusion among children and adults is due to how we define the word *food* and use the words *food* and *nutrients* interchangeably. Nutrients are substances that organisms require to carry out their life processes; they can be organic or inorganic. Not all nutrients provide energy. Examples of inorganic nutrients that do not provide energy are vitamins, minerals, and water. Examples of organic nutrients that provide energy are carbohydrates (including simple sugars), lipids (fats), and proteins. Food is a nutrient that contains energy, and it may contain inorganic

nutrients as well. Food provides energy and the building blocks for growth and tissue repair; it can be used immediately or stored for later use. For example, many plants store food in the form of starch.

All foods can be considered nutrients, but not all nutrients are considered food. To be considered food in a biological sense, the substance must contain energy that can be released during cellular respiration. Inorganic nutrients such as water and minerals are essential to metabolic processes but do not provide energy. The "plant food" commonly sold in stores is not food in a biological sense. It provides a source of inorganic nutrients that may not be present in the soil. Likewise, soil is not food but rather a source of plant nutrients such as minerals and water. Leaves are plant structures in which photosynthesis takes place and sugars are made. The leaves and other plant structures are then food for animals that eat plants. Sunlight is the form of energy used by the plant during photosynthesis, but it is not a substance and does not provide the building blocks needed to grow or repair plant structures. Chlorophyll is a substance contained in the plant's chloroplasts that is involved in photosynthesis. Fats, oils, and proteins are also foods. The only item on the list that is considered food for a plant is sugar.

Curricular and Instructional Considerations

Elementary Students

In the elementary grades students learn about

the needs of organisms. Through a variety of instructional opportunities, students learn that animals take their food in from the environment by eating plants, animals, or both. Students wonder about the differences between plants and animals and ask questions such as "How do plants get food?" (NRC 1996, p. 128). They learn that plants need nutrients and may be introduced to the idea that plants make their own food, but the ideas related to the process of photosynthesis are not developed until middle school. Elementary students also learn about food groups and nutrients in the context of human nutrition. Identifying sugar as the food plants use exceeds grade-level expectations for elementary students. However, the probe is useful in identifying ideas that form early about food for plants, particularly the notion that plants get their food from the soil.

Middle School Students

In middle school, students are introduced to the basic process of photosynthesis. They learn that plants make sugar from carbon dioxide and water using energy from sunlight and that the sugar can be used by the plant as a source of energy and as material for growth and repair or can be stored for later use. Middle school is the time when students need to develop a scientific conception of food different from the common, everyday use of the word. Even though the basic process of photosynthesis—including the idea that sugars are plants' *only* food— has been taught, middle school students may

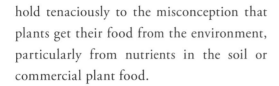

hold tenaciously to the misconception that plants get their food from the environment, particularly from nutrients in the soil or commercial plant food.

High School Students

In high school, students deepen their understanding by connecting their growing understanding of chemistry to the biological process of photosynthesis. The understanding that food is a source of energy is expanded to include the idea of energy from sunlight stored in the chemical bonds that form between the sugar's carbon atoms. This probe is useful at the high school level because it will often reveal that even after students have learned about photosynthesis, the idea that energy is released when chemical bonds are broken, and the biological concept of food, they may revert to their prior conceptions about where plants get their food.

Administering the Probe

This probe can be used with elementary students by removing unfamiliar words such as *chlorophyll* and substituting *air* for *carbon dioxide* and *oxygen*. For high school students the word *sugar* can be added to or substituted with *simple sugar* or *glucose*. High school teachers may also add *starch, oils,* and *protein* to the list. This probe may also be used as a card sort.

Related Ideas in *National Science Education Standards* (NRC 1996)

K–4 The Characteristics of Organisms

- Organisms have basic needs. For example, animals need air, water, and food; plants require air, water, nutrients, and light.

5–8 Populations and Ecosystems

- ★ For ecosystems, the major source of energy is sunlight. Energy entering ecosystems as sunlight is transferred by producers into chemical energy through photosynthesis.

9–12 The Cell

- ★ Plant cells contain chloroplasts, the site of photosynthesis. Plants and many microorganisms use solar energy to combine molecules of carbon dioxide and water into complex, energy-rich organic compounds and release oxygen to the environment. This process of photosynthesis provides a vital connection between the Sun and the energy needs of living systems.

9–12 Matter, Energy, and Organization in Living Systems

- ★ The energy for life primarily derives from the Sun. Plants capture energy by absorbing light and using it to form strong (covalent) chemical bonds between atoms of carbon containing (organic) molecules. These molecules can be used to assemble larger molecules with biological activity (including proteins, DNA, sugars, and fats). In addition, the energy stored in bonds between the atoms (chemical energy) can be used as sources of energy for life processes.

★ Indicates a strong match between the ideas elicited by the probe and a national standard's learning goal.

Related Ideas in *Benchmarks for Science Literacy* (AAAS 1993)

. . . .

K–2 Cells

- Most living things need food, water, and air.

K–2 Flow of Matter and Energy

- Plants and animals both need to take in water, and animals need to take in food. In addition, plants need light.

3–5 Flow of Matter and Energy

- Some source of "energy" is needed for all organisms to stay alive and grow.

6–8 Flow of Matter and Energy

★ Food provides molecules that serve as fuel and building material for all organisms. Plants use the energy in light to make sugars out of carbon dioxide and water. This food can be used immediately for fuel or materials, or it may be stored for later use.

Related Research

- Much of the research on students' ideas about food for plants was conducted in the 1980s and still applies to students' ideas today. Universally the most persistent notion is that plants take their food from the environment, particularly the soil. Students also believe that plants have multiple sources of food (Driver et al. 1994).

- Children appear to consider food as anything useful taken into an organism's body,

including water, minerals, and, in the case of plants, carbon dioxide or even sunlight. Typically, students do not consider starch as food for plants. Their reasoning is that starch is something plants make, not something they eat (Driver et al. 1994).

- Students often give a nonfunctional explanation about why plants and animals need food. They say it is needed to keep them alive, rather than describing the role of food in metabolism (Driver et al. 1994).

- In a study by Wandersee (1983) that surveyed 1,405 students ages 10–19 about the product of photosynthesis, most students selected proteins, relating them to food for growth, rather than energy. Some students in this study also mentioned plants getting vitamins from the soil.

- In a study by Tamir (1989), some students thought sunlight, associated with energy, was the food for plants. Many students also considered minerals taken in from the soil as food or believed that minerals had a direct role in photosynthesis.

- The everyday reference to fertilizers as "plant food" may promote the idea of fertilizer as being the food for plants (Driver et al. 1994).

- The idea that plants take their food in from the environment, rather than making it internally, is a common misconception that is highly resistant to change. Even when taught how plants make food by photosynthesis, students still hold on to the notion that food is taken in from the outside (AAAS 1993).

★ Indicates a strong match between the ideas elicited by the probe and a national standard's learning goal.

- Some children consider chlorophyll to be a food substance (Driver et al. 1994).

Suggestions for Instruction and Assessment

- Take the time to elicit students' definitions of the word *food*; many students use this word in a way that is not consistent with its biological meaning (AAAS 1993). Have students identify the difference between the everyday use of the word and the scientific use. Contrasting the two uses and providing examples may help them see the difference and begin to use the scientific definition.

- Understanding that the food plants make is very different from other nutrients such as water and minerals may be a prerequisite for understanding the idea that plants make their food rather than acquire it from the environment (Roth, Smith, and Anderson 1983).

- Many researchers note the conceptual demands of the topic of plant nutrition and point out that to understand the abstract and complex concept of photosynthesis, students need to possess the prerequisite concepts of living things, gas, food, and energy. Before introducing a chemical equation for photosynthesis, first help students understand that "an element, carbon (which is solid in its pure form), is present in carbon dioxide (which is a colorless gas in the air) and that this gas is converted by a green plant into sugar (a solid, but in solution) when hydrogen (a gas) from wa-

ter (a liquid) is added using light energy, which is consequently converted to chemical energy" (Driver et al. 1994, p. 30).

- High school students can often define photosynthesis and provide the equation, but questions that ask them to apply a basic understanding are often not asked of students. Use questions that encourage students to use the concept of photosynthesis to explain food, growth, and energy-related plant ideas.

- Show a container of plant food and a container of vitamins for humans. Build an analogy between the two to show that their purpose is to provide essential inorganic nutrients, not food energy.

Related NSTA Science Store Publications and Journal Articles

American Association for the Advancement of Science (AAAS). 1993. *Benchmarks for science literacy.* New York: Oxford University Press.

Driver, R., A. Squires, P. Rushworth, and V. Wood-Robinson. 1994. *Making sense of secondary science: Research into children's ideas.* London: RoutledgeFalmer.

George, R. 2003. How do plants make their own food? *Science & Children* (Feb.): 17.

Keeley, P. 2005. *Science curriculum topic study: Bridging the gap between standards and practice.* Thousand Oaks, CA: Corwin Press.

National Research Council (NRC). 1996. *National science education standards.* Washington, DC: National Academy Press.

Wali Abdi, S. 2006. Correcting student miscon-

ceptions. *Science Scope* 29 (4): 39.

Weinburgh, M. 2004. Teaching photosynthesis: More than a lecture but less than a lab. *Science Scope* 27 (9): 15–17.

References

American Association for the Advancement of Science (AAAS). 1993. *Benchmarks for science literacy.* New York: Oxford University Press.

Driver, R., A. Squires, P. Rushworth, and V. Wood-Robinson. 1994. *Making sense of secondary science: Research into children's ideas.* London: RoutledgeFalmer.

Keeley, P. 2005. *Science curriculum topic study: Bridging the gap between standards and practice.* Thousand Oaks, CA: Corwin Press.

National Research Council (NRC). 1996. *National science education standards.* Washington, DC: National Academy Press.

Roth, K., E. Smith, and C. Anderson. 1983. *Students' conceptions of photosynthesis and food for plants.* East Lansing, MI: Michigan State University, Institute for Research on Teaching.

Tamir, P. 1989. Some issues related to the justifications to multiple choice answers. *Journal of Biological Education* 11 (1): 48–56.

Wandersee, J. 1983. Students' misconceptions about photosynthesis: A cross-age study. In *Proceedings of the international seminar: Misconceptions in science and mathematics,* eds. H. Helm and J. Novak, 441–446. Ithaca, NY: Cornell University.

Giant Sequoia Tree

The giant sequoia tree is one of the largest trees on earth. It starts as a small seedling and grows into an enormous tree. Five children can stretch their arms across the width of the trunk of one of the large sequoia trees!

Where did most of the matter that makes up the wood and leaves of this huge tree originally come from? Circle the best answer.

A sunlight

B water

C soil

D carbon dioxide

E oxygen

F minerals

G chlorophyll

Explain your thinking. How did you decide where most of the matter that makes up this tree came from?

_____

_____

_____

_____

_____

_____

Giant Sequoia Tree

Teacher Notes

Purpose

The purpose of this assessment probe is to elicit students' ideas about transformation of matter. The probe is designed to reveal whether students recognize that a gas from the air (carbon dioxide) is combined with water and transformed into the new material that makes up most of the matter of the tree.

Related Concepts

photosynthesis, plants, transformation of matter

Explanation

The best response is D, carbon dioxide. Plants take in carbon dioxide (a gas) through their leaves and water from the soil and use the energy from sunlight to rearrange the atoms into new substances—sugar and oxygen. This process happens inside the leaf of the plant. Sunlight provides the energy for this process to happen. Chlorophyll is a pigment found within the leaf cells that absorbs the energy from sunlight used for the reaction. After food is made in the leaf, it travels to other parts of a plant, where it is used for energy, tissue repair, and growth or stored for later use.

Most of the matter, including the leaves and wood, that makes up the structure of the tree can be traced back to the carbon dioxide that was transformed into sugar through photosynthesis and used for building material. The mass contributed by the carbon dioxide is much greater than the mass contributed by the water. The atomic mass of one molecule of carbon dioxide is approximately 44 atomic mass units; one molecule of water is approximately 18 atomic mass units. When wood is burned, carbon dioxide and water vapor are released and go back into the air. When the wood is completely burned, the remaining ashes consist of the small amount of minerals taken in from the soil.

Curricular and Instructional Considerations

Elementary Students

In the elementary grades students learn that plants need sunlight, water, and nutrients to grow and stay healthy. Upper elementary students begin to recognize that plants make their own food. However, it is too abstract an idea for them to understand the transformation of matter that takes place during photosynthesis and growth of a plant. Both younger and older students have difficulty accepting the idea that something as seemingly light as air could make up the bulk weight or mass of a tree, partly because students lack opportuni-

ties to recognize that air is a substance that has weight (the term *mass* can be used with older elementary students). It is critical for students to have opportunities to accept the idea early on that air is matter and has mass (*weight* for younger students). This probe is useful in identifying early conceptions students have developed about where the material that makes up a tree came from.

Middle School Students

In middle school, students learn about chemical reactions and the types of transformations of matter that occur during these reactions. By the end of eighth grade they can begin to use the notion of atoms to explain what happens when matter is transformed in a process like photosynthesis. However, even though students can manipulate models to learn what happens during the transformation of carbon dioxide and water into sugar and oxygen, they may still refuse to recognize that a gas contributes the most mass to this reaction. It seems counterintuitive to students that most of the mass of the matter of a tree comes from carbon dioxide in the air. This probe is useful in determining whether students hold on to their intuitive idea about where most of the matter that makes up a plant comes from, even after instruction.

High School Students

In high school students learn more about the complex reaction of photosynthesis. Transformation of matter in a biological context now focuses more on the flow of matter through food webs. Students' increasing knowledge of chemistry, particularly carbon-based molecules, comes in handy when reasoning through a problem such as this one that involves molecular mass. When comparing molecular masses of carbon dioxide and water, carbon dioxide has more mass to contribute to the reaction. This probe is useful at the high school level because it will often reveal that even though students are taught photosynthesis and develop the idea that matter is transformed in a biological context, they may still revert to their prior conceptions about where most of the mass comes from based on their preconception that gases have negligible mass or that plants take their food in from the soil.

Administering the Probe

This probe can be used with upper elementary students by removing the choices *carbon dioxide* and *oxygen* and replacing them with *air*. The sequoia tree was used as the subject of this probe because of its massive size, but a more familiar tree may be substituted. Similar to the "seed and log" question in the Private Universe series (Harvard-Smithsonian Center for Astrophysics 1995), you might show a maple seed or acorn, a seedling of the tree, and a log cut from the tree and ask students where most of the "stuff" of the log came from as it grew from seed to seedling to large tree. The unscientific word *stuff* can be used intentionally in this probe to explore students' ideas without being hindered by their understanding of the concept of matter or mass.

Related Ideas in *National Science Education Standards* (NRC 1996)

K–4 The Characteristics of Organisms

*   Organisms have basic needs. For example, animals need air, water, and food; plants require air, water, nutrients, and light.

K–4 Properties of Objects and Materials

*   Objects are made up of one or more materials, such as paper, wood, and metal.

5–8 Structure and Function in Living Systems

★ Cells carry on the many functions needed to sustain life. They grow and divide, thereby producing more cells. This requires that they take in nutrients, which they use to provide energy for the work that cells do and to make the materials that a cell or organism needs.

5–8 Populations and Ecosystems

★ For ecosystems, the major source of energy is sunlight. Energy entering ecosystems as sunlight is transferred by producers into chemical energy through photosynthesis.

5–8 Properties and Changes in Properties of Matter

*   Substances react chemically in characteristic ways with other substances to form new substances (compounds) with different characteristic properties.

★ Indicates a strong match between the ideas elicited by the probe and a national standard's learning goal.

9–12 The Cell

★ Plant cells contain chloroplasts, the site of photosynthesis. Plants and many microorganisms use solar energy to combine molecules of carbon dioxide and water into complex, energy-rich organic compounds and release oxygen to the environment.

9–12 Matter, Energy, and Organization in Living Systems

★ The energy for life primarily derives from the Sun. Plants capture energy by absorbing light and using it to form strong (covalent) chemical bonds between atoms of carbon containing (organic) molecules. These molecules can be used to assemble larger molecules with biological activity (including proteins, DNA, sugars, and fats). In addition, the energy stored in bonds between the atoms (chemical energy) can be used as sources of energy for life processes.

9–12 Chemical Reactions

• Complex chemical reactions involving carbon-based molecules take place constantly in every cell in our bodies.

Related Ideas in *Benchmarks for Science Literacy* (AAAS 1993)

K–2 Flow of Matter and Energy

• Plants and animals both need to take in water, and animals need to take in food. In addition, plants need light.

K–2 The Structure of Matter

• Objects can be described in terms of the materials they are made of and their physical properties.

3–5 Flow of Matter and Energy

• Some source of "energy" is needed for all organisms to stay alive and grow.

3–5 The Structure of Matter

• When a new material is made by combining two or more materials, it has properties that are different from the original materials.

3–5 The Earth

• Air is a substance that surrounds us and takes up space.

6–8 Flow of Matter and Energy

★ Food provides molecules that serve as fuel and building material for all organisms. Plants use the energy in light to make sugars out of carbon dioxide and water. This food can be used immediately for fuel or materials, or it may be stored for later use.

• Energy can change from one form to another in living things. Animals get energy from oxidizing their food, releasing some of its energy as heat.

9–12 The Structure of Matter

• Atoms often join with one another in various combinations in distinct molecules or three-dimensional repeating patterns. An enormous variety of biological, chemical, and physical phenomena can be explained

★ Indicates a strong match between the ideas elicited by the probe and a national standard's learning goal.

by changes in the arrangement and motion of atoms and molecules.

Related Research

- The question in this probe is based on a similar question used in the Private Universe series where Harvard graduates were shown a seed and a log and asked where most of the mass of the log came from. Very few mentioned carbon dioxide. The most common responses were that it came from the soil or that it came from the water (Harvard-Smithsonian Center for Astrophysics 1995).
- Students have a difficult time imagining plants as chemical systems. In particular, middle school students think organisms and materials in the environment are very different types of matter. For example, plants are made of leaves, stems, and roots; the nonliving environment is made of water, soil, and air. Students see these substances as fundamentally different and not transformable into each other (AAAS 1993).
- Students have a difficult time accepting that weight increase and growth in plants is attributed to the incorporation of matter from a gas. In a study of 15-year-old students, many failed to mention carbon dioxide as the source of the increase in weight of growing seedlings, even though they knew that carbon dioxide was taken in during photosynthesis (Driver et al. 1994).
- Barker and Carr (1989) found that many children regarded sunlight as one of the reactants in photosynthesis, along with

carbon dioxide and water.

- Some students consider light to be made of molecules and thus contributing to the matter of a plant (Driver et al. 1994).
- Driver's study of 759 15-year-old students who had studied photosynthesis connected the idea of growth with photosynthesis. Although about a third of the students could understand the component idea of photosynthesis, only 8% could relate it to plant growth by describing how a tree makes tissue from the constituents it takes in from the environment. Only 3 students out of 759 said that tree tissue is made from carbon dioxide and water using light energy (Driver et al. 1994).
- In Wandersee's study (1983) of 1,405 students ages 10–19, many thought that the soil in a plant pot would lose weight as the plant grows because the plant uses the soil for food.

Suggestions for Instruction and Assessment

- Before students can accept the idea that the mass of a plant comes mostly from the carbon dioxide in the air, they have to accept air as matter that has weight or mass. Students need multiple opportunities to discover that gases have significant mass.
- High school students can use molecular masses to show that even though water is taken in and transformed along with carbon dioxide, the carbon dioxide molecules contribute significantly more mass than water molecules.

- Manipulating physical models of molecules may help middle school and high school students see what happens to the carbon dioxide.

- Photosynthesis is a complex reaction that is frequently treated as an equation with little opportunity to learn how the process contributes to the growth and energy needs of a plant. Explicitly make connections between the transformation of matter that occurs and plant growth.

- If students fail to recognize that carbon dioxide as a gas has weight, show students dry ice and explain that it is a solid form of carbon dioxide. Have students put on protective gloves and hold a piece of dry ice to sense the "felt weight."

- Use Von Helmont's experiment as a context to learn how the question of where plants got the materials they needed to grow from was historically explored. Have students evaluate the results of his experiment.

Related NSTA Science Store Publications and Journal Articles

American Association for the Advancement of Science (AAAS). 1993. *Benchmarks for science literacy.* New York: Oxford University Press.

Driver, R., A. Squires, P. Rushworth, and V. Wood-Robinson. 1994. *Making sense of secondary science: Research into children's ideas.* London: RoutledgeFalmer.

George, R. 2003. How do plants make their own food? *Science & Children* (Feb.): 17.

Keeley, P. 2005. *Science curriculum topic study:*

Bridging the gap between standards and practice. Thousand Oaks, CA: Corwin Press.

National Research Council (NRC). 1996. Photosynthesis vignette. In *National science education standards,* 194–196. Washington, DC: National Academy Press.

Weinburgh, M. 2004. Teaching photosynthesis: More than a lecture but less than a lab. *Science Scope* 27 (9): 15–17.

Related Curriculum Topic Study Guide

(Keeley 2005)
"Photosynthesis and Respiration"

References

American Association for the Advancement of Science (AAAS). 1993. *Benchmarks for science literacy.* New York: Oxford University Press.

Barker, M., and M. Carr. 1989. Photosynthesis: Can our pupils see the wood for the trees? *Journal of Biological Education* 23 (1): 41–44.

Driver, R., A. Squires, P. Rushworth, and V. Wood-Robinson. 1994. *Making sense of secondary science: Research into children's ideas.* London: RoutledgeFalmer.

Harvard-Smithsonian Center for Astrophysics. 1995. *Private Universe Project: Workshop 2. Biology: Why are some ideas so difficult?* Videotape. Burlington, VT: Annenberg/CPB Math and Science Collection.

Keeley, P. 2005. *Science curriculum topic study: Bridging the gap between standards and practice.*

Thousand Oaks, CA: Corwin Press.

National Research Council (NRC). 1996. *National science education standards*. Washington, DC: National Academy Press.

Wandersee, J. 1983. Students' misconceptions about photosynthesis: A cross-age study. In *Proceedings of the international seminar: Misconceptions in science and mathematics,* eds. H. Helm and J. Novak, 441–446. Ithaca, NY: Cornell University.

Baby Mice

Seif's pet mouse had babies. Five of the babies were black and two were white. The father mouse was black. The mother mouse was white. Seif and his friends wondered why the mice were different colors. These were their ideas:

Jerome: Baby mice inherit more traits from their fathers than their mothers.

Alexa: The baby mice got half their traits from their father and half from their mother.

June: Male traits are stronger than female traits.

Seif: Black mice have more traits than white mice.

Fiona: The black baby mice are probably male and the white baby mice are probably female.

Lydia: Parent's traits like fur color don't matter—nature decides what something will look like.

Billy: Blood type determines what traits babies will have.

Which friend do you most agree with and why? Explain your thinking.

_____

_____

_____

_____

_____

_____

Baby Mice

Teacher Notes

Purpose

The purpose of this assessment probe is to elicit students' basic ideas about inheritance of genetic traits. The probe is designed to reveal the variety of ideas students have about how traits, such as fur color, are passed on to offspring.

Related Concepts

chromosomes, genes, inherited traits

Explanation

The best response is Alexa's. The first step in the production of offspring from the two mice is fertilization of the female's egg by the male's sperm. Egg and sperm each contain half the number of mouse chromosomes. Genes are found on chromosomes. A gene is a segment of DNA on a chromosome that carries instructions for a particular trait, such as fur color. During fertilization, matched pairs of chromosomes, half from the mother and half from the father, come together and a single cell results, which will divide and eventually become the baby mouse. The baby mouse contains a full set of chromosomes—with half the genes coming from the mother and half from the father. The combination that results determines the offspring's characteristics.

One way in which genes are expressed was described by Gregor Mendel, who believed that traits could be either dominant or recessive. When two genes for the same trait are paired and one of the genes is dominant, the dominant gene will be expressed. In the example of the mice, black fur color is dominant. Even if the offspring only have one gene for black fur, the trait that will be expressed is black fur. White fur is a recessive trait that is expressed when a dominant gene is not present. The white offspring have two genes for white fur color. Mendelian genetics is a first step in understanding how genes are expressed, but understanding genetics is much more complex.

The key idea in this probe is that an organism's inherited traits are determined by the pairing of genes from the mother and father, with each parent contributing 50% of the genes; the combination of dominant and recessive genes determines which traits are expressed. It is not the result of one sex having more or stronger traits (or genes) as described in Jerome's and June's responses. Black and white mice have the same number of genes (contrary to Seif's response); they are just expressed differently. Coat color in mice is not determined by sex as described in Fiona's response. For example, some of the white mice could be male if they received a recessive gene from both the mother and father. Lydia's response targets the idea of acquired characteristics, which are not inherited from an organism's parents. For example, if a mouse lost its tail in an accident, that would be an acquired trait—it would not be passed on to the offspring. Lydia's response is a teleological argument that implies that some intentional force of nature directs the traits that offspring will exhibit, rather than that traits are the result of gene expression. Billy's response is similar to historical beliefs. Before Mendel, many people thought traits were passed on through the blood.

Curricular and Instructional Considerations

Elementary Students

In the elementary grades, students are just beginning to learn about inherited characteristics. They observe that offspring do not always look exactly like their parents or each other. In the later elementary grades they begin to develop an understanding that traits are passed on from parents to offspring, but it is too early to introduce the mechanism of inheritance and the role of genes. By eliminating some of the distracters, this probe can be used to examine students' early ideas about how traits are passed on to offspring.

Middle School Students

In middle school, students learn basic ideas about the mechanism of inheritance, combining ideas about reproduction, cell division, and basic genetics. They develop an understanding of the role of chromosomes and genes in passing on characteristics from one generation to the next. The expectation at this grade level is that students should understand that half of their genes come from their mother and half from their father. This combination results in the inherited traits they exhibit. Students should recognize the role of chance in determining which chromosomal pairs come together during fertilization and that probability can help predict the outcome of inherited characteristics. However, the detailed mechanism of genetics exceeds the middle school level. This probe is useful in identifying whether students have preconceived ideas about how genetic traits are passed on to offspring.

High School Students

In high school, students learn the details of Mendelian genetics and how various gene combinations occur and express themselves. They should be able to explain why some traits are expressed and some are not. They learn about genetics at a molecular level, including the role of DNA. This probe is useful at the high school level in determining students' precursor ideas before planning and teaching a unit on genetics.

Administering the Probe

This probe is best administered to middle and high school students. If using the probe with elementary students, substitute *characteristics* for *traits* if they are unfamiliar with the latter term. Depending on when genes are introduced in the curriculum, upper middle school or high school teachers can substitute the term *genes* for *traits* in Jerome's, Alexa's, June's, and Seif's responses.

Related Ideas in *National Science Education Standards* (NRC 1996)

· ·

K–4 Life Cycles of Organisms

• Plants and animals closely resemble their parents.

• Many characteristics of an organism are inherited from the parents of the organism, but other characteristics result from an individual's interactions with the environment. Inherited characteristics include the color of flowers and the number of limbs of an animal.

5–8 Reproduction and Heredity

★ In many species, including humans, females produce eggs and males produce sperm. Plants also produce sexually—the egg and sperm are produced in the flowers of flowering plants. An egg and a sperm unite to begin development of a new individual. That new individual receives genetic information from its mother (via the egg) and its father (via the sperm). Sexually produced offspring are never identical to either of their parents.

★ Every organism requires a set of instructions for specifying its traits. Heredity is the passage of these instructions from one generation to another.

• Heredity information is contained in genes, located in the chromosomes of each cell. Each gene carries a single unit of information.

9–12 The Molecular Basis of Heredity

• In all organisms, the instructions for specifying the characteristics of the organism are carried in DNA.

★ Transmission of genetic information to offspring occurs through egg and sperm cells that contain only one representative from each chromosome pair. An egg and a sperm unite to form a new individual.

Related Ideas in *Benchmarks for Science Literacy* (AAAS 1993)

· ·

★ Indicates a strong match between the ideas elicited by the probe and a national standard's learning goal.

K–2 Heredity

- There is variation among individuals of one kind within a population.

3–5 Heredity

- For offspring to resemble their parents, there must be a reliable way to transfer information from one generation to the next.

6–8 Heredity

- ★ In some kinds of organisms, all the genes come from a single parent, whereas in organisms that have sexes, typically half of the genes come from each parent.

9–12 Heredity

- ★ The sorting and recombination of genes in sexual reproduction results in a great variety of possible gene combinations from the offspring of any two parents.

Related Research

- "Early middle-school students explain inheritance only in observable features, but upper middle-school and high-school students have some understanding that characteristics are determined by a particular genetic entity which carries information translatable by the cell" (AAAS 1993, p. 341).

- When asked to describe how physical traits are passed from parents to offspring, elementary, middle, and high school students all exhibited misconceptions, including the idea that traits are inherited from only one of the parents and that certain traits only come from the mother or father (AAAS

1993).

- Studies have shown that it may not be until middle school that students can include in their explanations of inheritance the role of chance and probability (AAAS 1993).

- In a study by Hackling and Treagust (1982), 94% of 15-year-old students understood the concept that one's characteristics come from parents, 50% understood that reproduction and inheritance occur together, and 44% understood that one gets a mixture of features from both parents (Driver et al. 1994).

- In a sample of 52 students ages 11–14, Deadman and Kelly (1978) found that boys had a prevalent conception that characteristics from male parents were stronger in their expression (Driver et al. 1994).

- Engel Clough and Wood-Robinson (1985) found that some students had a tendency to favor the mother as the primary contributor of genetic material as well as a belief that daughters inherit from mothers and sons inherit from fathers (Driver et al. 1994).

- In a study of ideas about the mechanism of inheritance among children ages 7–13, Kargbo, Hobbs, and Erickson (1980) found that half the children gave a naturalistic explanation, such as nature makes offspring resemble parents. Some thought traits were decided by the brain or blood. Only a few children in the sample, who were among the older children in the group, mentioned any genetic principle. In analyzing the students' responses, the authors found that they were not giving flippant, unconsidered answers

★ Indicates a strong match between the ideas elicited by the probe and a national standard's learning goal.

but rather were drawing on established frameworks to make sense of inheritance (Driver et al. 1994).

- Several researchers have found that even before students receive formal instruction in genetics, they know the word *gene* and, less frequently, *chromosome*. However, students have little understanding of the nature or function of genes or chromosomes (Driver et al. 1994).

- Research shows that students have some idea of the randomness of inheritance, meaning that sometimes offspring are like their mother, sometimes they are like their father, and sometimes they are like both. However, students rarely show evidence of applying the concepts of chance and probability to common situations even after advanced courses (Driver et al. 1994).

- Use of the word *dominant* in regard to dominant and recessive traits may contribute to several misconceptions. For example, students may think that dominant traits are "stronger" and "overpower" the recessive trait, that dominant traits are more likely to be inherited, that dominant traits are more prevalent in the population, that dominant traits are "better," and that male or masculine traits are dominant (Donovan 1997).

Suggestions for Instruction and Assessment

- Children in early grades should have observational experiences to compare how offspring of familiar animals resemble each other and their parents, describing and drawing examples of similarities and differences.

- In upper elementary grades, rather than describing a characteristic (e.g., the mouse has black fur) students should begin to develop an inventory of traits that come from parents (e.g., fur color). They should discuss and have opportunities to resolve differences in opinion about traits that come from parents, traits that come from interaction with the environment, characteristics that are learned, and things they are unsure about.

- In middle school, combine the study of genetics with the study of reproduction.

- Use caution with terminology at the high school level when teaching genetics, particularly with the concept of dominance so as not to imply the idea that some genes are "stronger" than others.

- Genetics terminology may hinder conceptual learning when terms are used imprecisely. If students are told "Inherited traits are carried on chromosomes," they may then confuse the terms *trait* and *gene*. (Genes, not traits, are carried on chromosomes.) Clear and consistent use of terms such as *trait, gene,* and *allele* is essential for constructing an accurate conceptual foundation of genetics (Bryant 2003).

- Caution should be used when students are asked to develop or use models to represent the mechanism of inheritance. Some models oversimplify the process of random assortment, recombination, and pairing of genes and expression of traits.

- Be aware of problems in using Punnett

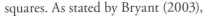

squares. As stated by Bryant (2003),

"The Punnett square works well for studying the inheritance of genetic traits controlled by a single gene, and can even be applied when two or more traits are considered simultaneously, as long as the genes are not located on the same chromosome (linked). Students often learn to use Punnett squares to obtain correct answers to genetics problems, but they fail to understand that a Punnett square represents two biological processes—gamete formation and fertilization. Students rely on Punnett squares as algorithms for getting the "right answer," often at the expense of meaningful conceptual understanding" (p. 11).

Related NSTA Science Store Publications and Journal Articles

American Association for the Advancement of Science (AAAS). 2001. *Atlas of science literacy.* (See "DNA and Inherited Characteristics," pp. 68–69.) New York: Oxford University Press.

Baker, W., and C. Thomas. 1998. Gummy bear genetics. *The Science Teacher* 65 (8): 25.

Bryant, R. J. 2003. Toothpick chromosomes: Simple manipulatives to help students understand genetics. *Science Scope* 26 (7): 10–15.

Driver, R., A. Squires, P. Rushworth, and V. Wood-Robinson. 1994. *Making sense of secondary science: Research into children's ideas.* London: RoutledgeFalmer.

Fetters, M. K., and M. Templin. 2002. Building traits. *The Science Teacher* 69 (4): 56–60.

Hazen, R., and J. Trefil. 1992. *Science matters.* (See "Code of Life," pp. 224–242.) New York: Anchor Books.

Keeley, P. 2005. *Science curriculum topic study: Bridging the gap between standards and practice.* Thousand Oaks, CA: Corwin Press.

National Research Council (NRC). 1996. *National science education standards.* Washington, DC: National Academy Press.

Rice, E., M. Krasny, and M. Smith. 2006. *Garden genetics: Teaching with edible plants.* Arlington, VA: NSTA Press.

Related Curriculum Topic Study Guides

(Keeley 2005)
"Mechanism of Inheritance (Genetics)"
"Reproduction, Growth, and Development (Life Cycles)"

References

American Association for the Advancement of Science (AAAS). 1993. *Benchmarks for science literacy.* New York: Oxford University Press.

Bryant, R. J. 2003. Toothpick chromosomes: Simple manipulatives to help students understand genetics. *Science Scope* 26 (7): 10–15.

Deadman, J., and P. Kelly. 1978. What do secondary school boys understand about evolution and heredity before they are taught the topics? *Journal of Biological Education* 12 (1): 7–15.

Donovan, M. 1997. The vocabulary of biology and the problem of semantics. *Journal of College Science Teaching* 26: 381–382.

Driver, R., A. Squires, P. Rushworth, and V. Wood-Robinson. 1994. *Making sense of secondary science: Research into children's ideas.* London: RoutledgeFalmer.

Engel Clough, E., and C. Wood-Robinson. 1985. Children's understanding of inheritance. *Journal of Biological Education* 19 (4): 304–310.

Hackling, M., and D. Treagust. 1982. What lower secondary students should understand about the mechanisms of inheritance, and what they should do following instruction. *Research in Science Education* 12: 78–88.

Kargbo, D., E. Hobbs, and G. Erickson. 1980. Children's beliefs about inherited characteristics. *Journal of Biological Education* 14 (2): 137–146.

Keeley, P. 2005. *Science curriculum topic study: Bridging the gap between standards and practice.* Thousand Oaks, CA: Corwin Press.

National Research Council (NRC). 1996. *National science education standards.* Washington, DC: National Academy Press.

Whale and Shrew

The blue whale is the largest mammal in the world. The pygmy shrew is one of the smallest mammals in the world. How does the size of average cells compare between a blue whale and a pygmy shrew? Circle the answer that best matches your thinking.

A The average cell of a blue whale is smaller than the average cell of a pygmy shrew.

B The average cell of a blue whale is larger than the average cell of a pygmy shrew.

C The average cell of a blue whale is about the same size as the average cell of a pygmy shrew.

Explain your thinking. Describe the "rule" or reasoning you used to choose your answer.

_____

_____

_____

_____

_____

_____

_____

_____

_____

Whale and Shrew

Teacher Notes

Purpose

The purpose of this assessment probe is to elicit students' ideas about cell size. The probe is designed to find out if students think that animal cell size is related to the overall size of an animal.

Related Concepts

cells, growth, cell division

Explanation

The best answer is C: The average cell of a blue whale is about the same size as the average cell of a pygmy shrew. The size of average mam-mal cells (this excludes cells that are unusually large, such as neurons) is similar in all mammal species. Even though some body cells (such as neurons) can be very large and cells vary, the average body cells of most mammals are about 10 micrometers in diameter. Interestingly, the earliest-stage embryos of the whale and shrew are also a similar size, even though a whale eventually reaches a mass of 150,000 kg whereas a mouse only reaches 15g—a 10-million-fold difference!

Cells are limited in how large they can be because the surface area-to-volume ratio does not stay the same as the size of a cell increases. Cells need to be able to move materials into and out of a cell, and it is harder for a large cell to pass materials in and out of the membrane and to move materials through the cell. The reason blue whales are larger than pygmy shrews is because they have more cells, not because their cells are larger.

Curricular and Instructional Considerations

Elementary Students

In the upper elementary grades students are just beginning to learn about cells as the fundamental unit of all living organisms. They observe a variety of cells of single-celled and multicelled organisms in pictures and with simple microscopes. At this level, students are not ready to compare cell sizes, but they do observe that larger animals have larger body parts such as legs and teeth as well as larger organs such as heart, lungs, and stomach. This can lead to a preconception that their cells are also larger. Students learn about growth in the context of life cycles but do not yet equate growth with an increase in the number of cells. However, this probe is useful in determining if the idea of cell size increasing with overall animal size and size of body parts is a conception that develops early on.

Middle School Students

Middle school students extend their observations of cells to making comparisons of similar cell types across animal species. Students develop the idea of similarities among species by

examining internal structures as well as cells. They can also begin to recognize the very small size of most cells and that most cells repeatedly divide to make more cells. Organisms and the organs they contain generally grow in size from birth until they reach adulthood. Yet, students may believe that the cells that make up organs are proportional to the size of the organ and thus the size of the animal, not recognizing that it is the process of cell division that contributes to growth, not the individual cells getting larger.

High School Students

High school students have a deeper understanding of the cell, including cell division and what controls it. Mathematically they develop an understanding of the relationship between volume and surface area and how the total surface area decreases with an increase in volume. Through lab experiences with model cells made of gels, they observe how the surface area-to-volume ratio affects the passage of materials into, around, and out of a cell, thus limiting the size of a cell. This probe is useful in eliciting students' ideas before designing experiences that help students understand that cell size is limited by the surface area-to-volume ratio and thus is relatively similar for most mammals' cells.

Administering the Probe

Show a picture of a whale and a shrew to contrast size. Make sure middle and high school students focus on the concept of "average-

sized" cells. You might explain that some cells, such as neurons, vary considerably in size. If necessary, choose a particular cell, such as a red blood cell or a cell from the liver. Alert students to the fact that the picture on the probe handout is not drawn to scale. In a scale drawing, the shrew would be several times smaller than the size of the whale's eye and would be a mere dot next to the whale.

Related Ideas in *National Science Education Standards* (NRC 1996)

. . . .

5–8 Structure and Function in Living Systems

• All organisms are composed of cells—the fundamental unit of life.

★ Cells carry on the many functions needed to sustain life. They grow and divide, thereby producing more cells.

5–8 Diversity and Adaptations of Organisms

★ Millions of species of animals, plants, and microorganisms are alive today. Although different species might look dissimilar, the unity among organisms becomes apparent from an analysis of internal structures, the similarity of their chemical processes, and evidence of common ancestry.

9–12 The Cell

• Cells can differentiate, and complex multicellular organisms are formed as a highly organized arrangement of differentiated

cells. In the development of these multicellular organisms, the progeny from a single cell form an embryo in which the cells multiply and differentiate to form the many specialized cells, tissues, and organs that comprise the final organism.

Related Ideas in *Benchmarks for Science Literacy* (AAAS 1993)

. . . .

3–5 Cells

• Microscopes make it possible to see that living things are made mostly of cells. Some organisms are made of a collection of similar cells that benefit from cooperating.

6–8 Cells

• All living things are composed of cells, from just one to many millions, whose details usually are visible only through a microscope. Different body tissues and organs are made up of different kinds of cells. The cells in similar tissues and organs in other animals are similar to those in human beings but differ somewhat from cells found in plants.

★ Cells continually divide to make more cells for growth and repair.

Related Research

• Stavy and Tirosh (2000) asked students in grades 7–12 a question similar to the one in this probe, comparing muscle cells of a mouse to muscle cells of an elephant. The majority of students, especially in grades 7

★ Indicates a strong match between the ideas elicited by the probe and a national standard's learning goal.

and 8, thought that larger animals have larger cells. The common justification was that "according to the dimensions of the elephants and those of the mice, it is obvious that the muscle cells of the mice are smaller than those of the elephants" (p. 30). This is an example of the intuitive rule "more A, more B." Most of the younger students who answered correctly explained the equality in terms of the cells having the same function and therefore being the same size. Most of the high school students who responded correctly used formal biological knowledge of cells and also described the elephant as having more cells.

- Available research on cells is limited. However, in piloting this probe with more than 100 middle and high school students, many students chose answer B (the blue whale has larger cells than a shrew). Their reasoning matched Stavy and Tirosh's results and was based on the idea that whales are much larger and therefore need larger cells.

Suggestions for Instruction and Assessment

- When students are examining the same cell types of different organisms, encourage them to look not only at the similarity in the shape of the cells but also the size. For example, when comparing the muscle cells of frogs to the blood cells of humans, notice the similar size.
- Develop the idea that cell size is limited by

the ability of molecules to pass in, around, and out of cells. Older students can test this idea by making model cells out of blocks of agar of different volume to surface area-to-volume ratios and measuring the rate and depth of penetration of a dye into the model cell. Calculate the volume to surface area ratios of the different cell sizes and compare the results of the diffusion based on the ratios.

- Have students investigate the question, "Is bigger always better?" in the context of a cell's ability to carry out its life functions. Encourage them to develop a way to research and test their idea and have them share their results.
- Ask students why a paramecium can never be the size of human. Develop the idea of why single-celled organisms must be microscopic to carry out the same life processes carried out by multicellular organisms.

Related NSTA Science Store Publications and Journal Articles

American Association for the Advancement of Science (AAAS). 1993. *Benchmarks for science literacy.* New York: Oxford University Press.

Driver, R., A. Squires, P. Rushworth, and V. Wood-Robinson. 1994. *Making sense of secondary science: Research into children's ideas.* London: RoutledgeFalmer.

Keeley, P. 2005. *Science curriculum topic study: Bridging the gap between standards and practice.* Thousand Oaks, CA: Corwin Press.

National Research Council (NRC). 1996. *National science education standards.* Washington, DC: National Academy Press.

Rau, G. 2004. How small is a cell? *The Science Teacher* (Oct.): 38–41.

Related Curriculum Topic Study Guides

(Keeley 2005)

"Cells"

References

American Association for the Advancement of Science (AAAS). 1993. *Benchmarks for science literacy.* New York: Oxford University Press.

Keeley, P. 2005. *Science curriculum topic study: Bridging the gap between standards and practice.* Thousand Oaks, CA: Corwin Press.

National Research Council (NRC). 1996. *National science education standards.* Washington, DC: National Academy Press.

Stavy, R., and D. Tirosh. 2000. *How students (mis-) understand science and mathematics: Intuitive rules.* New York: Teachers College Press.

Habitat Change

A small, short-furred, gray animal called a divo lives on an island. This island is the only place on Earth where divos live. The island habitat is warm and provides plenty of the divos' only food—tree ants. The divos live high in the treetops, hidden from predators.

One year the habitat experienced a drastic change that lasted for most of the year. It became very cold and even snowed. All of the ants died. The trees lost their leaves, but plenty of seeds and dried leaves were on the ground.

Circle any of the things you think happened to most of the divos living on the island after their habitat changed.

A The divos' fur grew longer and thicker.

B The divos switched to eating seeds.

C The divos dug holes to live under the leaves or beneath rocks.

D The divos hibernated through the cold period until the habitat was warm again.

E The divos died.

Explain your thinking. How did you decide what effect the change in habitat would have on most of the divos?

_____

_____

_____

_____

Habitat Change

Teacher Notes

Purpose

The purpose of this assessment probe is to elicit students' ideas about adaptation. The probe is designed to reveal whether students think individuals intentionally change their physical characteristics or behaviors in response to an environmental change.

Related Concepts

adaptation, behavioral response

Explanation

There is no one completely right answer to this question, but the best answer is E: The divos died. In common usage the term *adapt* is un-

derstood as any type of change over any span of time. Individuals generally do not intentionally adapt to drastic changes in their environment by changing their physical characteristics (such as fur length or ability to eat certain foods based on teeth or mouth structure) or inherited behaviors (such as where they seek shelter or hibernation). Some individual divos may have been born with variations that made them better suited to survive a change in the environment and to reproduce, passing on their traits to new generations that would be better adapted to the changed environment. However, most of the divos probably died because the physical structures, physiology, and behaviors they were born with no longer fit the

changed environment. Populations can adapt over time, but individuals do not change during their lifetimes.

Curricular and Instructional Considerations

Elementary Students

In the elementary grades students build understandings of biological concepts through direct experience with living things and their habitats. The idea that organisms depend on their environment is not well developed in young children. The focus in the early elementary grades should be on establishing the primary association of organisms with their environments, followed by the upper elementary ideas of dependence on various aspects of the environment and structures and behaviors animals were born with that help various organisms survive (NRC 1996). Students should have opportunities to investigate a variety of habitats of plants and animals and identify ways animals and plants depend on the environment and each other.

Middle School Students

Understanding adaptation can be particularly troublesome at this level. Many students think *adaptation* means that individuals change in major ways in response to environmental changes (i.e., if the environment changes, individual organisms deliberately adapt) (NRC 1996, p. 156). Teachers need to carefully select activities that do not imply to students that an individual organism can change its

structures and behaviors at will when the habitat changes. Students at this level should develop the idea of variations organisms are born with that can lead to an individual's survival and reproduction.

High School Students

At the high school level students shift from thinking about the selection of individuals with certain traits that help them survive in their environment to the changing proportion of such traits in a population of organisms. The idea of natural selection, leading to the culminating idea of biological evolution, is a major focus in biology. However, some students still believe that a change in an organism's structures or behaviors in response to its environment can be controlled by the organism and passed along to future generations.

Administering the Probe

Explain to students that the divo is an imaginary organism. However, the challenges it faces, due to the drastic change in its environment, would produce similar responses from real organisms. Consider adding additional distracters for structural changes (such as growing stronger teeth for cracking open seeds) or behavioral changes (such as learning to swim so it could get off the island).

Related Ideas in *National Science Education Standards* (NRC 1996)

K–4 The Characteristics of Organisms

- Organisms have basic needs. For example, animals need air, water, and food; plants require air, water, nutrients, and light. Organisms can survive only in environments where their needs can be met.

K–4 Organisms and Their Environments

- An organism's patterns of behavior are related to the nature of that organism's environment, including the kinds and numbers of other organisms present, the availability of food and resources, and the physical characteristics of the environment. When the environment changes, some plants and animals survive and reproduce, and others die or move to new locations.

5–8 Regulation and Behavior

- All organisms must be able to obtain and use resources, grow, reproduce, and maintain stable internal conditions while living in a constantly changing external environment.
- Behavior is one kind of response an organism can make to an internal or environmental stimulus. A behavioral response requires coordination and communication at many levels, including cells, organ systems, and whole organisms. Behavioral response is a set of actions determined in part by heredity and in part from experience.

5–8 Diversity and Adaptations of Organisms

- ★ Species acquire many of their unique characteristics through biological adaptation, which involves the selection of naturally occurring variations in populations. Biological adaptations include changes in structures, behaviors, or physiology that enhance survival and reproductive success in a particular environment.
- ★ Extinction of a species occurs when the environment changes and the adaptive characteristics of a species are insufficient to ensure its survival.

Related Ideas in *Benchmarks for Science Literacy* (AAAS 1993)

K–2 Diversity of Life

- Plants and animals have features that help them live in different environments.

K–2 Heredity

- There is variation among individuals of one kind within a population.

3–5 Interdependence of Life

- For any particular environment, some kinds of plants and animals survive well, some survive less well, and some cannot survive at all.
- ★ Changes in an organism's habitat are sometimes beneficial to it and sometimes harmful.

6–8 Interdependence of Life

- ★ In any particular environment, the growth and survival of organisms depend on the physical conditions.

★ Indicates a strong match between the ideas elicited by the probe and a national standard's learning goal.

9–12 Diversity of Life

- The variation of organisms within a species increases the likelihood that at least some members of the species will survive under changed environmental conditions.

9–12 Heredity

- An altered gene may be passed on to every cell that develops from it. The resulting features may help, harm, or have little effect on the offspring's success in its environment.

Related Research

- Many students tend to see adaptation as an intention by the organism to satisfy a desire or need for survival (Driver et al. 1994).

- Middle school and high school students may believe that organisms are able to intentionally change their bodily structure to be able to live in a particular habitat or that they respond to a changed environment by seeking a more favorable environment. It has been suggested that the language about adaptation used by teachers or textbooks may cause or reinforce these beliefs (AAAS 1993, p. 342).

- Most students seem to think that adaptation involves individual organisms changing in major ways as a response to survive a change in their environment (Driver et al. 1994).

- Students appear to confuse an individual's adaptation during its lifetime and inherited changes in a population over time. A

large number of students appear to adopt a Lamarckian view of adaptation (Driver et al. 1994).

Suggestions for Instruction and Assessment

- Some "adaptations" are controlled by an organism. When dealing with individual organisms, *acclimatization* would be a better term to use for noninheritable changes in structure or behavior made by an organism during its lifetime.

- Be aware that Lamarckian interpretations of an individual's adaptation to its environment may impede understanding of Darwinian evolution.

- A common activity used in elementary and middle school science is to have students design an imaginary organism that is adapted to a particular habitat. Be aware that this activity may perpetuate the misconception that organisms intentionally adapt.

- Compare and contrast with students the everyday common use of the word *adaptation* with the scientific meaning of the word. Add this to students' growing number of examples of the ways we use words in our society that are not always the same as the scientific use of the words.

Related NSTA Science Store Publications and Journal Articles

American Association for the Advancement of Science (AAAS). 1993. *Benchmarks for science lit-*

★ Indicates a strong match between the ideas elicited by the probe and a national standard's learning goal.

eracy. New York: Oxford University Press.

Driver, R., A. Squires, P. Rushworth, and V. Wood-Robinson. 1994. *Making sense of secondary science: Research into children's ideas.* London: RoutledgeFalmer.

Endreny, A. 2006. Crazy about crayfish. *Science & Children* 43 (7): 32–35. Also available online at www.nsta.org/main/news/stories/science_and_children.php?news_story_ID=51806.

Keeley, P. 2005. *Science curriculum topic study: Bridging the gap between standards and practice.* Thousand Oaks, CA: Corwin Press.

National Research Council (NRC). 1996. *National science education standards.* Washington, DC: National Academy Press.

Royce, C. 2005. Antarctic adaptations. *Science & Children* (Jan.): 16–18.

Young, J. 2005. Feeding of *Diarmis proboscis. Science Scope* (Jan.): 56–58.

Related Curriculum Topic Study Guides
(Keeley 2005)
"Adaptation"
"Habitats and Local Environments"

References

American Association for the Advancement of Science (AAAS). 1993. *Benchmarks for science literacy.* New York: Oxford University Press.

Driver, R., A. Squires, P. Rushworth, and V. Wood-Robinson. 1994. *Making sense of secondary science: Research into children's ideas.* London: RoutledgeFalmer.

Keeley, P. 2005. *Science curriculum topic study: Bridging the gap between standards and practice.* Thousand Oaks, CA: Corwin Press.

National Research Council (NRC). 1996. *National science education standards.* Washington, DC: National Academy Press.

Earth and Space Science Assessment Probes

Earth and Space Science Assessment
Concept Matrix

Probes

Core Science Concepts	Geology — Is It a Rock? (Version 1)	Geology — Is It a Rock? (Version 2)	Geology — Mountaintop Fossil	Day/Night Cycle — Darkness at Night	Sky Objects — Emmy's Moon and Stars	Sky Objects — Objects in the Sky
Day/Night Cycle				✓		
Earth's Axis				✓	✓	✓
Fossils			✓			
Landforms			✓			
Light Reflection						✓
Light Source						✓
Minerals	✓	✓				
Rocks	✓	✓				
Rotation				✓		
Scale Size and Distance in the Universe					✓	✓
Stars					✓	✓
Uplift			✓			
Weathering and Erosion	✓		✓			

Is It a Rock? (Version 1)

Which things on this list could be rocks? How do you decide if something is a rock? Put a X next to the things you think could be a rock.

___ jagged boulder   ___ smooth boulder

___ small stone   ___ large stone

___ pebble   ___ piece of gravel

___ piece of sand   ___ dust from two stones rubbed together

Explain your thinking. What "rule" or reasoning did you use to decide if something is a rock?

_____

_____

_____

_____

_____

_____

_____

_____

_____

_____

Is It a Rock? (Version 1)

Teacher Notes

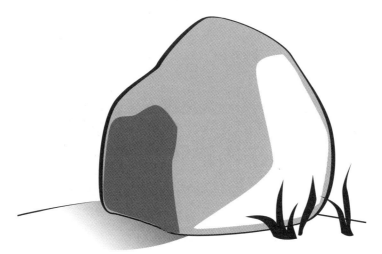

Purpose

The purpose of this assessment probe is to elicit students' ideas about rocks. The probe is designed to determine whether students recognize that rocks can come in many sizes and shapes.

Related Concepts

minerals, rocks, weathering and erosion

Explanation

All of the items on the list could be a rock. Rocks are aggregates of minerals. Simply, a *rock* is defined as any solid mass of mineral or mineral-like matter that occurs naturally as part of our planet (Lutgens and Tarbuck 2003). Rocks can be described by their size and shape. Rocks can range from huge boulders to single grains of sand and rock dust

formed through the process of weathering. They can be jagged or smooth. Words like *boulder, gravel,* and *sand* have specific scientific meanings related to the average size of rock fragments. Rocks can be broken and shaped by natural weathering processes or broken, cut, and shaped by humans, resulting in a variety of sizes and shapes. Some items on this list can also be minerals; for example, a grain of sand could be the mineral quartz as well as a grain of volcanic rock. However, this probe focuses only on the concept of a rock.

Curricular and Instructional Considerations

Elementary Students

Younger elementary students should become familiar with their immediate surroundings, including the variety of rocks in their local environment. Students should observe the different shapes and sizes rocks come in. In upper elementary grades students can observe details of rocks and minerals and can use magnifiers to identify grains of rock and minerals in sand. At this level, students begin to understand that smooth rocks are shaped by the action of waves, wind, and water and that freezing water and other natural processes as well as human actions can break rock up into smaller pieces, including rock dust. Upper elementary students begin to develop an understanding that the solid materials formed by the Earth are rocks, minerals, and soil. When they investigate soils, they begin to develop the idea that small particles of rock are combined with living and once-living material.

Middle School Students

Students continue to refine their ideas about how rocks are formed, shaped, and broken apart by the action of both abrupt and slow natural processes. They begin to tie these processes to the idea of a rock cycle. They develop an understanding that sediments contain small particles of rock and minerals and that these sediments can be cemented together again to form solid rock. They develop an understanding of how landforms such as mountains are formed through the uplift of rock or the hardening of molten lava from volcanoes and how these landforms break down into rock of different sizes and shapes, including grains of sand found on beaches made from volcanic rock.

High School Students

Students at this level refine their understanding of the rock cycle—the formation, weathering, sedimentation, and reformation through heat and/or pressure, resulting in large rock formations as well as particles of rock of different sizes and shapes at different stages of the rock cycle. Nevertheless, some students may still hold on to their preconceptions that rocks are defined by size or shape.

Administering the Probe

Make sure younger students are familiar with the objects and materials on the list. Remove unfamiliar examples. It may be helpful to have props that show items on the list, including a picture of a large boulder. This probe can also be administered as a card sort. Place the words and/or pictures of the items on cards and ask students to sort them by "rock," "not rock," or "unsure" and to provide an explanation for each one.

Related Ideas in *National Science Education Standards* (NRC 1996)

K–4 Properties of Earth Materials

- Earth materials are solid rocks and soils, water, and the gases of the atmosphere. The varied materials have different physical and chemical properties.

5–8 Structure of the Earth System

- Some changes in the solid Earth can be described as the "rock cycle." Old rocks at the

Earth's surface weather, forming sediments that are buried, then compacted, heated, and often crystallized into new rock. Eventually, those new rocks may be brought to the surface by forces that drive plate motions, and the rock cycle continues.

Related Ideas in *Benchmarks for Science Literacy* (AAAS 1998)

K–2 Processes That Shape the Earth

★ Chunks of rocks come in many sizes and shapes, from boulders to grains of sand and even smaller.

3–5 Processes That Shape the Earth

- Rock is composed of different combinations of minerals. Smaller rocks come from the breakage and weathering of bedrock and larger rocks. Soil is made partly from weathered rock, partly from plant remains—and also contains many living organisms.

- Waves, wind, water, and ice shape and reshape the Earth's land surface by eroding rock and soil in some areas and depositing them in other areas, sometimes in seasonal layers.

6–8 Processes That Shape the Earth

- Sedimentary rock buried deep enough may be reformed by pressure and heat, perhaps melting and recrystallizing into different kinds of rock. These reformed rock layers may be forced up again to become land

surface and even mountains. Subsequently, this new rock will erode.

9–12 Processes That Shape the Earth

- The formation, weathering, sedimentation, and reformation of rock constitute a continuing "rock cycle" in which the total amount of material stays the same as its form changes.

Related Research

- Freyberg (1985) found that the word *rock* is used in many different ways in our common language, contributing to the confusion over what a rock is geologically. Many students think rocks are of a particular size rather than characterized by what they are made of (Driver et al. 1994).

- Children often fail to recognize that words like *boulder, gravel, sand,* and *clay* have specific meanings related to the average size of fragments. For example, children think of clay as being sticky, orange stuff found underground rather than a very fine particle of rock (Driver et al. 1994).

- A study by Happs (1982) revealed that younger students often intuitively identify rocks through their weight, hardness, color, and jaggedness. Therefore, some students believe that rocks are larger, heavier, and jagged and identify smaller fragments as stones instead of rocks (Driver et al. 1994).

- Students have difficulty with the idea of rock types being a range of sizes. They use the words *boulder, gravel, sand,* and *clay* in

---

★ Indicates a strong match between the ideas elicited by the probe and a national standard's learning goal.

ways related to where they are found, rather than seeing them as rocks of different sizes. For example, they say that boulders are larger than rocks and have rolled down a slope, gravel is something on the side of roads, sand is on beach and in the desert, and clay is red and underground (Happs 1985).

Suggestions for Instruction and Assessment

- Younger students should be given ample opportunities to collect and examine a variety of rocks of different sizes and describe them according to their observable properties.

- When younger elementary students describe physical properties of objects, include rocks in their study of properties of matter. Have students identify that rocks come in different sizes, shapes, colors, and textures.

- Provide opportunities for students to see that rocks can break down into very small pieces, including "rock dust." This can be observed by rubbing two rocks together or filling a clean coffee can with a few rocks and shaking it for an extended period of time. The dust comes from the pieces of weathered rock. Ask students what the dust is; encourage them to make the connection that it came from the rock and thus is the same material.

- Compare and contrast different types of beach sand formed from rocks and minerals and trace their origin. For example, the quartz in some sands may have come from

pieces of granite rock that were further weathered into mineral particles. Volcanic beach sand comes from the weathering of volcanic rock.

- Throughout grades K–12 students have many opportunities to learn about rocks, their properties, and the processes that formed them. Yet students' conceptions of what a rock actually is may be nothing more than a memorized definition. Encourage students to develop an operational definition of a rock, and help them bridge their operational definition to a scientific one through multiple experiences.

Related NSTA Science Store Publications and NSTA Journal Articles

American Association for the Advancement of Science (AAAS). 1993. *Benchmarks for science literacy.* New York: Oxford University Press.

American Association for the Advancement of Science (AAAS). 2001. *Atlas of science literacy* (See "Changes in the Earth's Surface," pp. 50–51.) New York: Oxford University Press.

Damonte, K. 2004. Going through changes. *Science & Children* (Oct.): 25–26.

Driver, R., A. Squires, P. Rushworth, and V. Wood-Robinson. 1994. *Making sense of secondary science: Research into children's ideas.* London: RoutledgeFalmer.

Ford, B. 1996. *Project earth science: Geology.* Arlington, VA: NSTA Press.

Keeley, P. (2005). *Science curriculum topic study: Bridging the gap between standards practice.* Thousand Oaks, CA: Corwin Press.

National Research Council (NRC). 1996. *National science education standards.* Washington, DC: National Academy Press.

Plummer, D., and W. Kulman. 2005. Rocks in our pockets. *Science Scope* 29 (2): 60–61.

Varelas, M., and J. Benhart. 2004. Welcome to rock day. *Science & Children* 40 (1): 40–45.

Related Curriculum Topic Study Guides

(Keeley 2005)

"Rocks and Minerals"

"Processes That Change the Surface of the Earth"

"Weathering and Erosion"

References

American Association for the Advancement of Science (AAAS). 1993. *Benchmarks for science literacy.* New York: Oxford University Press.

Driver, R., A. Squires, P. Rushworth, and V. Wood-Robinson. 1994. *Making sense of secondary science: Research into children's ideas.* London: RoutledgeFalmer

Freyberg, P. 1985. Implications across the curriculum. In *Learning in science,* eds. R. Osborne and P. Freyberg, 125–135. Auckland, New Zealand: Heinemann.

Happs, J. 1982. *Rocks and minerals.* LISP Working Paper 204. Hamilton, New Zealand. University of Waikato, Science Education Research Unit.

Happs, J. 1985. Regression in learning outcomes: Some examples from Earth science. *European Journal of Science Education* 7 (4): 431–443.

Keeley, P. 2005. *Science curriculum topic study: Bridging the gap between standards and practice.* Thousand Oaks, CA: Corwin Press.

Lutgens, F., and E. Tarbuck. 2003. *Essentials of geology.* 8th ed. Upper Saddle River, NJ: Prentice Hall.

National Research Council (NRC). 1996. *National science education standards.* Washington, DC: National Academy Press.

Is It a Rock? (Version 2)

What is a rock? How do you decide if something is a rock?
Put an X next to the things that you think are rocks.

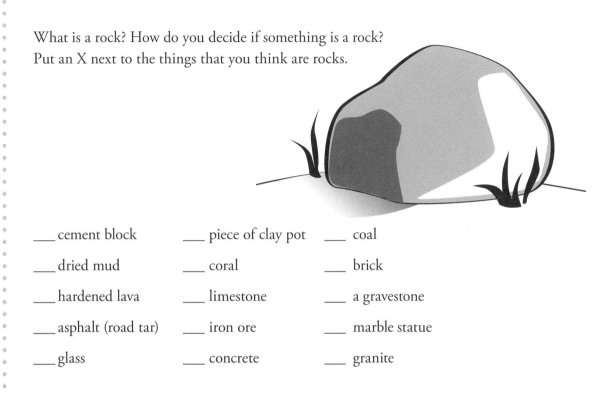

___ cement block   ___ piece of clay pot   ___ coal

___ dried mud   ___ coral   ___ brick

___ hardened lava   ___ limestone   ___ a gravestone

___ asphalt (road tar)   ___ iron ore   ___ marble statue

___ glass   ___ concrete   ___ granite

Explain your thinking. What "rule" or reasoning did you use to decide if something
is a rock?

_____

_____

_____

_____

_____

_____

_____

_____

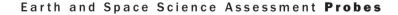

Is It a Rock? (Version 2)

Teacher Notes

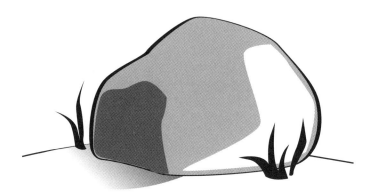

Purpose

The purpose of this assessment probe is to elicit students' ideas about rocks. The probe is designed to determine whether students can distinguish between human-made, "rock-like" materials and geologically formed rock material of various origins, even though it may have been shaped by humans. The probe reveals whether students have a geologic conception of a rock.

Related Concepts

minerals, rocks

Explanation

The items on the list that are rocks are coal, hardened lava, limestone, a gravestone, iron ore, marble statue, and granite. Simply, a rock can be defined as any solid mass of mineral or mineral-like matter that occurs naturally as part of our planet (Lutgens and Tarbuck 2003).

Some rocks, such as limestone, are composed almost entirely of one mineral—in this case, impure masses of calcite. Other rocks occur as aggregates of two or more minerals. For example, granite is a common rock composed of the minerals quartz, hornblende, and feldspar. A few rocks are composed of nonmineral matter. Pumice is a volcanic rock formed by the cooling of frothy lava. Coal is a rock formed by the hardening of solid organic debris.

Some of the items on the list are rock-like in that they are similar to rock material but are not naturally formed through geologic processes. The cement block, piece of clay pot, brick, asphalt, glass, and concrete are all made using some rock material, combined with other materials, and reshaped through a human-made process, not a geologic one. The material itself is not "rock." However, the gravestone and marble statue are rock, even though they

have been reshaped and polished through a human-made process, because the material they are made of was formed through a geologic process and the original composition is unchanged. The material is still rock, only the shape and texture have changed.

Coral is made by living processes, not geologic processes. Soft-bodied organisms secrete calcium carbonate to make hard, rock-like casings that protect their soft bodies. These "community casings" result in the formation of coral reefs.

Mud is a mixture of silt, clay, and water. Silt and clay are fine rock fragments. Mud can dry out, forming hard cakes that appear rock-like. However, it takes long periods of geologic time for dried mud to harden (lithify) into solid sedimentary rock such as shale.

Curricular and Instructional Considerations

Elementary Students

Observing and classifying objects and materials is a major part of elementary science inquiry. Younger elementary students should become familiar with the variety of objects and materials in their local environment, including rocks and objects made from rocks. Ideas about rocks are linked to ideas about properties of matter. They begin to understand that rocks can come in natural forms or can be cut, shaped, and polished by humans for various uses. They begin to understand how some objects and materials exist naturally and others are made by humans

combining materials from the environment in new ways, based on the properties of the materials. Students compare and classify familiar human and naturally made objects and materials, and they investigate unfamiliar materials to find out what they are made of.

Middle School Students

Students continue to refine their ideas about how natural objects such as rocks are formed. They can contrast composition and formation of human-made materials with naturally formed ones. They begin to develop an understanding of how rocks are formed through various geologic processes, resulting in a variety of sedimentary, igneous, and metamorphic rocks. Students can begin to trace the composition of rocks and minerals back to the geologic processes that formed them. They can contrast this formation with short-term human processes developed through materials science and technology that result in rock-like materials such as cement. In their study of natural resources, they recognize that rock is a natural resource that can be reshaped by humans without changing its composition or can be crushed and combined with other materials to form a new, hard material.

High School Students

Student at this level refine their understanding of the geologic processes that form rocks as well as an understanding of the chemical composition and origin of minerals that make up rocks.

They have a greater awareness of the long-term, geologic processes that form rocks. They learn about chemical processes, invented by humans, which result in rock-like mixtures such as asphalt, concrete, and cement. In biology they recognize living processes that form hard, rock-like casings such as coral and mollusk shells and link this to the idea of biogeochemical cycles. At this level, combined with their knowledge of chemistry, students have greater familiarity with synthetically produced materials and are more apt to differentiate them from materials produced through geologic processes.

Administering the Probe

Make sure younger students understand the words on the list. You may choose to show examples (actual or photographic) of the materials or point out examples they are familiar with in their local environment, such as a cement sidewalk. Words can also be written on cards or combined with pictures and used as a card sort activity, sorting cards into "rock" and "non-rock." For older students who are familiar with examples of igneous rocks, consider replacing the term *hardened lava* with *basalt* or *pumice*.

Related Ideas in *National Science Education Standards* (NRC 1996)

. . . .

K–4 Properties of Earth Materials

★ Earth materials are solid rocks and soils, water, and the gases of the atmosphere. The varied materials have different physical and chemical properties, which make them useful in different ways, for example, as building materials, as sources of fuel, or for growing the plants we use as food. Earth materials provide many of the resources that humans use.

K–4 Types of Resources

• Some resources are basic materials such as air, water, and soil; some are produced from basic resources, such as food, fuel, and building materials.

5–8 Structure of the Earth System

• Some changes in the solid Earth can be described as the "rock cycle." Old rocks at the Earth's surface weather, forming sediments that are buried, then compacted, heated, and often crystallized into new rock. Eventually, those new rocks may be brought to the surface by forces that drive plate motions, and the rock cycle continues.

9–12 Geochemical Cycles

• Each element on Earth moves among reservoirs in the solid Earth, oceans, atmosphere, and organisms as part of geochemical cycles.

9–12 Natural Resources

• Human populations use resources in the environment in order to maintain and improve their existence.

Related Ideas in *Benchmarks for Science Literacy* (AAAS 1998)

. . . .

⊠ Indicates a strong match between the ideas elicited by the probe and a national standard's learning goal.

K–2 The Structure of Matter

★ Objects can be described in terms of the materials they are made up of.

3–5 Processes That Shape the Earth

★ Rock is composed of different combinations of minerals.

3–5 Materials and Manufacturing

• Through science and technology, a wide variety of materials that do not appear in nature at all have become available.

6–8 Processes That Shape the Earth

★ Sedimentary rock buried deep enough may be reformed by pressure and heat, perhaps melting and recrystallizing into different kinds of rock. These reformed rock layers may be forced up again to become land surface and even mountains. Subsequently, this new rock will erode. Rock bears evidence of the minerals, temperatures, and forces that created it.

9–12 Processes That Shape the Earth

• The formation, weathering, sedimentation, and reformation of rock constitute a continuing "rock cycle" in which the total amount of material stays the same as its form changes.

Related Research

• The word *rock* is used in many different ways in our common language, contributing to the confusion of what it means in a geologic sense (Freyberg 1985).

• Some students regard rock as being made of only one substance and thus have difficulty in recognizing granite as rock (Driver et al. 1994).

• In studies by Happs (1982, 1985), students had difficulty making the distinction between "natural" things and those created or altered by humans. For example, some students considered brick a rock because part of it comes from natural material. Conversely, some students thought cut, smooth, polished marble is not a rock because humans made it smooth and so it is no longer natural (Driver et al. 1994).

• When students were shown different types of rock and asked whether they were rocks, several students thought pumice was too light to be a rock (Osborne and Freyberg 1985).

Suggestions for Instruction and Assessment

• When teaching about rocks, take time to elicit students' conception of what a rock is. Although students may have had several opportunities to study rocks during their K–8 experiences, do not assume that they have a correct conception of what a rock is. Students may be able to define a rock, name types of rocks, and describe the geologic processes that formed them, yet they may still identify human-made materials, such as brick, as rocks.

• Develop an operational definition before introducing the scientific definition. Students need to understand what minerals are before they develop a scientific notion

---

★ Indicates a strong match between the ideas elicited by the probe and a national standard's learning goal.

of rocks, based on their composition.

- Emphasize the long periods of geologic time it takes to make rock and review the stages of the rock cycle versus the short period of time to make a brick.

- When younger elementary students describe physical properties of objects, include rocks in the study of properties. Rocks can be used to demonstrate how a physical property may change, but the material is still the same. For example, show students a rough piece of granite and a smooth, polished piece of granite, noting that they are still the same material although the property of texture has been changed by humans.

- Compare and contrast naturally formed objects with objects made or reformed by humans. In the latter category have students place objects into two groups: (1) those made entirely from natural materials that have not been recombined when reshaped by humans (e.g., the marble statue) and (2) those that contain some natural material, combined with other materials to make new material that does not exist in a natural state (e.g., concrete or brick).

- Have students investigate the materials that make up brick, concrete, cement, and asphalt. Connect this to materials science and technology, noting how humans use natural resources and scientific knowledge about materials to make new types of materials for construction.

Related NSTA Science Store Publications and NSTA Journal Articles

American Association for the Advancement of Science (AAAS). 1993. *Benchmarks for science literacy.* New York: Oxford University Press.

Damonte, K. 2004. Going through changes. *Science and Children* (Oct.): 25–26.

Driver, R., A. Squires, P. Rushworth, and V. Wood-Robinson. 1994. *Making sense of secondary science: Research into children's ideas.* London: RoutledgeFalmer

Ford, B. 1996. *Project Earth science: Geology.* Arlington, VA: NSTA Press.

Keeley, P. 2005. *Science curriculum topic study: Bridging the gap between standards and practice.* Thousand Oaks, CA: Corwin Press.

National Research Council (NRC). 1996. *National science education standards.* Washington, DC: National Academy Press.

Plummer, D., and W. Kulman. 2005. Rocks in our pockets. *Science Scope* 29 (2): 60–61.

Varelas, M., and J. Benhart. 2004. Welcome to rock day. *Science & Children* 40 (1): 40–45.

Related Curriculum Topic Study Guide
(Keeley 2005)
"Rocks and Minerals"

References

American Association for the Advancement of Science (AAAS). 1993. *Benchmarks for science literacy.* New York: Oxford University Press.

Driver, R., A. Squires, P. Rushworth, and V. Wood-Robinson. 1994. *Making sense of secondary science: Research into children's ideas.* London: RoutledgeFalmer.

Freyberg, P. 1985. Implications across the curriculum. In *Learning in science,* eds. R. Osborne and P. Freyberg, 125–135. Auckland, New Zealand: Heinemann.

Happs, J. 1982. *Rocks and minerals.* LISP Working Paper 204. Hamilton, New Zealand. University of Waikato, Science Education Research Unit.

Happs, J. 1985. Regression in learning outcomes: Some examples from Earth science. *European Journal of Science Education* 7 (4): 431–443.

Keeley, P. (2005). *Science curriculum topic study: Bridging the gap between standards and practice.* Thousand Oaks, CA: Corwin Press.

Lutgens, F., and E. Tarbuck. 2003. *Essentials of geology.* 8th ed. Upper Saddle River, NJ: Prentice Hall.

National Research Council (NRC). 1996. *National science education standards.* Washington, DC: National Academy Press.

Osborne, R., and P. Freyberg. 1985. *Learning in science: The implications of children's science.* Auckland, New Zealand: Heinemann.

Mountaintop Fossil

The Esposito family went hiking on a tall mountain. Mrs. Esposito picked up a shell fossil on the top of the mountain. The fossil was once a shelled organism that lived in the ocean. The family had different ideas about how the fossil ended up there. This is what they thought:

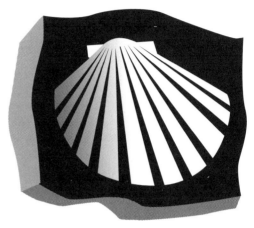

Mrs. Esposito: A bird picked up the organism and dropped the shell as it flew over the mountain.

Mr. Esposito: Water, ice, or wind eventually carried the fossil to the top of the mountain.

Rosa: A mountain formed in an area that was once covered by ocean.

Sofia: The fossil flowed out of a volcano that rose up from the ocean floor.

Whose idea do you most agree with and why? Describe your ideas about how a fossil could end up on the top of a tall mountain.

_____

_____

_____

_____

_____

_____

_____

_____

_____

Mountaintop Fossil

Teacher Notes

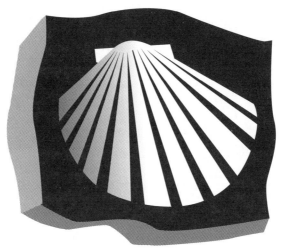

Purpose

The purpose of this assessment probe is to elicit students' ideas about mountain formation. The probe is designed to determine whether students recognize that mountains formed from the uplift of land, including areas that were once part of oceans.

Related Concepts

fossils, landforms, uplift, weathering and erosion

Explanation

The best answer is Rosa's. Over long periods of geologic time, the Earth's crust goes through several changes. Where oceans, shallow seas, and muddy marshes once existed, today there may be mountains. Ancient marine organisms died and were covered with sediments that, over time, hardened and formed sedimentary rock. The imprints left by the hard shells of mollusks and even mineralized parts of their shells remained in the sedimentary rock. Additional layers of sedimentary rock formed over the fossils. Over time, these layers of rock were uplifted to form mountains. As mountains formed, the fossils were elevated along with the rock in which they were formed. Today the processes of weathering and erosion expose the fossils in the rock that were formed millions of years ago. The sediments may end up in the ocean and again turn to rock over long periods of time, possibly forming new mountains many millions of years from now.

Curricular and Instructional Considerations

Elementary Students

Students at this level should have the opportunity to examine different types of landforms and rocks, including fossils, with the emphasis on observations and descriptions. Explanations should be based on processes and changes that students can experience and observe. It is difficult for students at this level to comprehend the very long periods of geologic time it takes for rocks and mountains to form.

Middle School Students

It is important for students at this age to understand how sedimentary rock is formed, including the embedding of plant and animal remains that leave a record of the appearance and disappearance of different species and the environment that existed at that time. The study of the Earth's history provides evidence about the evolution of the Earth's features, including the distribution of land and sea, features of the crust such as mountains, and the populations of living organisms that existed at different times. Students should have opportunities to study a variety of landforms, including mountains, and how they came to be. They should understand that the Earth has gone through many changes and that where oceans once existed, mountains may exist today. The theory of plate tectonics is introduced at this level.

Because students do not have direct contact with the phenomena of uplift and wearing down of mountains as well as the long-term nature of geologic processes, instruction and hands-on experiences should be descriptive. Detailed explanations should wait until late in middle school or high school. It is also important to note that vast intervals of geologic time are difficult for students to comprehend.

High School Students

At this level students transition from descriptive understandings of geologic phenomena they learned about in middle school to modern explanations, including plate tectonics. They should have an integrated knowledge about the Earth system that includes the rock cycle, crustal dynamics, geochemical processes, and the expanded concept of geologic time. They should understand and use the evidence base for determining the story of the Earth's crust, climate, and evolving life forms.

Administering the Probe

It may be helpful to show students an example of a shell fossil. You might also show a picture of a tall mountain chain, such as the Andes, where shell fossils have been found.

Related Ideas in *National Science Education Standards* (NRC 1996)

K–4 Properties of Earth Materials

★ Fossils provide evidence about plants and animals that lived a long time ago and the nature of the environment at that time.

5–8 Structure of the Earth

★ Landforms are the result of a combination of constructive and destructive forces.

★ Indicates a strong match between the ideas elicited by the probe and a national standard's learning goal.

Constructive forces include crustal deformation, volcanic eruption, and deposition of sediment; destructive forces include weathering and erosion.

5–8 Earth's History

- The Earth processes we see today, including erosion, movement of lithospheric plates, and changes in the atmospheric composition, are similar to those that occurred in the past. Earth history is also influenced by occasional catastrophes, such as the impact of an asteroid or comet.

9–12 The Origin and Evolution of the Earth System

- Geologic time can be estimated by observing rock sequences and using fossils to correlate the sequences at various locations. Current methods include using known decay rates of radioactive isotopes present in the rocks to measure time since the rock was formed.

- Interactions among the solid Earth, the oceans, the atmosphere, and organisms have resulted in the ongoing evolution of the Earth system. We can observe changes such as earthquakes and volcanic eruptions on a human time scale, but many processes such as mountain building and plate movements take place over hundreds of millions of years.

Related Ideas in *Benchmarks for Science Literacy* (AAAS 1998)

3–5 Processes That Shape the Earth

- Rock is composed of different combinations of minerals. Smaller rocks come from the breakage and weathering of bedrock and larger rocks. Soil is made partly from weathered rock, partly from plant remains—and also contains many living organisms.

6–8 Processes That Shape the Earth

- Sediments of sand and smaller particles (sometimes containing the remains of organisms) are gradually buried and cemented together by dissolved minerals and form solid rock again.

- Sedimentary rock buried deep enough may be reformed by pressure and heat, perhaps melting and recrystallizing into different kinds of rock. These reformed rock layers may be forced up again to become land surface and even mountains. Rock bears evidence of the minerals, temperatures, and forces that created it.

Related Research

- Students of all ages may hold the view that the world has always been the way it is now and any changes that occurred were sudden and comprehensive (Freyberg 1985).

- Very few younger children who were interviewed in a study by Happs (1982) appreciated the relationship between sedimentary rocks and the sedimentary process by which they were formed.

- Students often think of mountain building as occurring only through catastrophic

★ Indicates a strong match between the ideas elicited by the probe and a national standard's learning goal.

events such as earthquakes or volcanoes. They often fail to recognize the slow process of uplift over millions of years (Phillips 1991).

- Some students have a landform and ocean basin conception that involves a progressively decreasing slope from the center of the continents to the center of the bottom of the ocean and then back up again (Marques and Thompson 1997).

Suggestions for Instruction and Assessment

- Provide examples of tall mountains such as the Himalayas and the Andes and show examples of marine fossils that have been found there. Encourage students to think of all the possible ways these fossils could have gotten there and have them research their ideas.
- Students should see as many different types of landforms as possible to help determine and describe the different ways in which they formed.
- Elementary students can observe the basic processes of the rock cycle—weathering, erosion, transport, and deposit—using water, sandboxes, and rock tumblers. Later they can connect these experiences to explanations of how features of different Earth formations came to be and how they are always changing.
- Films or internet simulations of mountain-building processes, particularly the slower

uplifts and not the catastrophic types such as volcanoes, provide a vicarious way for students to observe long-term constructive processes.

Related NSTA Science Store Publications and NSTA Journal Articles

American Association for the Advancement of Science (AAAS). 1993. *Benchmarks for science literacy.* New York: Oxford University Press.

Driver, R., A. Squires, P. Rushworth, and V. Wood-Robinson. 1994. *Making sense of secondary science: Research into children's ideas.* London: RoutledgeFalmer.

Ford, B. 1996. *Project Earth science: Geology.* Arlington, VA: NSTA Press.

Gilbert, S., and S. Ireton. 2003. *Understanding models in earth and space science.* Arlington, VA: NSTA Press.

Hemler, D., and T. Repine. 2002. Reconstructing the geologic timeline: Adding a constructivist slant to a classic activity. *The Science Teacher* 69 (4): 32–35.

Keeley, P. 2005. *Science curriculum topic study: Bridging the gap between standards and practice.* Thousand Oaks, CA: Corwin Press.

National Research Council (NRC). 1996. *National science education standards.* Washington, DC: National Academy Press.

Norell, M. 2003. What is a fossil? *Science & Children* 40 (5): 20.

Phillips, W. 1991. Earth science misconceptions. *The Science Teacher* 58 (2): 21–23.

Related Curriculum Topic Study Guides

(Keeley 2005)

"Fossil Evidence"

"Landforms"

"Plate Tectonics"

"Processes That Change the Surface of the Earth"

"Rocks and Minerals"

"Weathering and Erosion"

References

American Association for the Advancement of Science (AAAS). 1993. *Benchmarks for science literacy.* New York: Oxford University Press.

Driver, R., A. Squires, P. Rushworth, and V. Wood-Robinson. 1994. *Making sense of secondary science: Research into children's ideas.* London: RoutledgeFalmer.

Freyberg, P. 1985. Implications across the curriculum. In *Learning in science,* eds. R. Osborne and P. Freyberg, 125–135. Auckland, New Zealand: Heinemann.

Happs, J. 1982. *Rocks and minerals.* LISP Working Paper 204. Hamilton, New Zealand. University of Waikato, Science Education Research Unit.

Keeley, P. 2005. *Science curriculum topic study: Bridging the gap between standards and practice.* Thousand Oaks, CA: Corwin Press.

Marques, L., and D. Thompson. 1997. Misconceptions and conceptual changes concerning continental drift and plate tectonics among Portuguese students aged 16-17. *Research in Science and Technological Education* 15 (2): 195–222.

National Research Council (NRC). 1996. *National science education standards.* Washington, DC: National Academy Press.

Phillips, W. 1991. Earth science misconceptions. *The Science Teacher* 58 (2): 21–23.

Darkness at Night

Six friends were wondering why the sky is dark at night. This is what they said:

Jeb: "The clouds come in at night and cover the Sun."

Talia: "The Earth spins completely around once a day."

Nick: "The Sun moves around the Earth once a day."

Becca: "The Earth moves around the Sun once a day."

Latisha: "The Sun moves underneath the Earth at night."

Yolanda: "The Sun stops shining."

Which friend do you think has the best reason for why the sky is dark at night? Describe your ideas about why the Earth is dark at night and light during the day.

_____

_____

_____

_____

_____

_____

_____

_____

Darkness at Night

Teacher Notes

Purpose

The purpose of this probe is to elicit students' ideas about the day/night cycle. The probe is designed to find out if students recognize that the Earth's rotation is responsible for the day/night cycle.

Related Concepts

day/night cycle, Earth's axis, rotation

Explanation

The best response is Talia's: The Earth spins completely around once a day. The reason for the day/night cycle is that the Earth spins completely around on its axis approximately every 24 hours. When our location on the Earth is turned away from the Sun, we have night (darkness). When our location on the

Earth is turned toward the Sun, we have day (daylight).

Curricular and Instructional Considerations

Elementary Students

In the early elementary years, students recognize that there is a repeating pattern of daytime and night. This is primarily observational; young children are not ready to explain the abstract idea of a turning Earth in relation to the Sun. Observations of the Sun's location make it look to them as if the Sun is the body that is moving. They may also have difficulty with the notion of a spherical Earth at this age, which is a precursor to understanding that the Earth spins.

In the upper elementary grades students begin to learn how the Earth moves in relation to the Sun. Using physical models such as globes and a light source, they begin to develop the notion that the Earth turns completely around on its axis once a day. Using models, students can see how people on the Earth experience darkness when they are turned away from the Sun. Gradually the terms *rotation* and *Earth's axis* are introduced when students are ready to conceptualize them. The ideas targeted by this probe are a grade-level expectation in the national standards.

Middle School Students

Students' understanding of the day/night cycle expands to include ideas about the effect of the tilt and the amount of sunlight that falls on the Earth at different times of the year. Students begin to recognize that the length of day (photoperiod) changes during different times of the year and different locations on the Earth, but the orbital geometry involved in understanding this concept is still difficult for them to grasp. Students at this level may begin to confuse rotation with revolution when the idea of the Earth's annual orbit around the Sun is introduced in middle school.

High School Students

During high school, more complex ideas about the Earth, Moon, and Sun system are developed along with the idea that other planets and their moons rotate at different speeds and have day/night cycles of varying lengths. How-

ever, be aware that high school students may not understand the basic concept of rotation. Some high school students still struggle with the concept of a rotating Earth in relation to the Sun as the mechanism that accounts for day/night.

Administering the Probe

Be sure students understand what the probe is asking. Terminology like *rotation, spinning on an axis,* and *revolution* are intentionally avoided to probe for conceptual understanding. Make sure students understand that *night* refers to the period of darkness when the Sun is not visible and that *day* refers to the period of daylight, not a 24-hour measurement.

Related Ideas in *National Science Education Standards* (NRC 1996)

· ·

K–4 Objects in the Sky

- The Sun, Moon, stars, clouds, birds, and airplanes all have properties, locations, and movements that can be observed.

K–4 Changes in the Earth and Sky

- Objects in the sky have patterns of movement. The Sun, for example, appears to move across the sky in the same way every day. But its path changes slowly over the seasons.

5–8 Earth in the Solar System

★ Most objects in the solar system are in

★ Indicates a strong match between the ideas elicited by the probe and a national standard's learning goal.

regular and predictable motion. Those motions explain such phenomena as the day, the year, phases of the Moon, and eclipses.

Related Ideas in *Benchmarks for Science Literacy* (AAAS 1993)

K–2 The Universe

- The Sun can be seen only in the daytime, but the Moon can be seen sometimes at night and sometimes during the day. The Sun, Moon, and stars all appear to move slowly across the sky.

3–5 The Earth

★ Like all planets and stars, the Earth is approximately spherical in shape. The rotation of the Earth on its axis every 24 hours produces the day/night cycle. To people on the Earth, this turning of the planet makes it seem as though the Sun, Moon, planets, and stars are orbiting the Earth once a day.

6–8 The Earth

- Because the Earth turns daily on an axis that is tilted relative to the plane of the Earth's yearly orbit around the Sun, sunlight falls more intensely on different parts of the Earth during the year.

Related Research

- Baxter (1989) identified six ideas about day and night and their prevalence and showed that students seems to move through these ideas as they get older: the Sun goes behind hills, clouds cover the Sun, the Moon covers the Sun, the Sun goes behind the Earth once a day, the Earth goes around the Sun once a day, and the Earth spins on it axis once a day. It appears that at ages 15 and 16 many still hold covering and orbital theories of day and night (Driver et al. 1994).

- Students at the secondary level may believe that day and night occur because the Earth goes around the Sun or the Sun goes around the Earth (Schoon 1992).

- Explanations of the day/night cycle, the phases of the Moon, and the seasons are very challenging for students. To understand these phenomena, students first should master the idea of a spherical Earth, itself a challenging task (AAAS 1993).

- Sadler (1987) noted that students may not be able to understand explanations of the day/night cycle, phases of the Moon, and seasons before they reasonably understand the relative size, motion, and distance of the Sun, Moon, and Earth (AAAS 1993).

Suggestions for Instruction and Assessment

- Make sure young students accept the idea of a spherical Earth, a precursor to understanding the spin of the Earth. Young students should also develop the idea of a repeated cycle of day/night before being expected to explain it.

- Use physical models made from common objects, such as a globe and a flashlight, to help students see the phenomenon of day and night. Encourage students to manipu-

---

★ Indicates a strong match between the ideas elicited by the probe and a national standard's learning goal.

late models rather than be passive observers.

• Students often confuse the terminology related to Earth's motion. Introduce *rotation* before *revolution,* starting with the concept before introducing the terminology. Once students have grasped the idea of rotation, use a model to help them see that the Earth rotates as it orbits around the Sun.

• Help students recognize how language may lead to incorrect ideas about the day/night cycle. Have students critique the use of words and phrases like *sundown, sunrise, the Sun is sinking,* and *the Sun is moving across the sky*; ask them to explain how these words or phrases may convey incorrect ideas about the Sun-Earth system. Have students demonstrate with models what is actually happening with the Sun in relation to the Earth.

• Students may know that the Earth spins on its imaginary axis, but they may have never been asked to describe what direction the Earth spins in—clockwise or counterclockwise, east to west or west to east? Challenge students to figure it out based on their observations of sunrise and sunset.

Related NSTA Science Store Publications and NSTA Journal Articles

American Association for the Advancement of Science (AAAS). 1993. *Benchmarks for science literacy.* New York: Oxford University Press.

Bogan, D., and D. Wood. 1997. Simulating Sun, Moon, and Earth patterns. *Science Scope* 21 (2): 46, 48.

Driver, R., A. Squires, P. Rushworth, and V. Wood Robinson. 1994. *Making sense of secondary science: Research into children's ideas.* London: RoutledgeFalmer.

Gilbert, S., and S. Ireton. 2003. *Understanding models in Earth and space science.* Arlington, VA: NSTA Press.

Keeley, P. 2005. *Science curriculum topic study: Bridging the gap between standards and practice.* Thousand Oaks, CA: Corwin Press.

National Research Council (NRC). 1996. *National science education standards.* Washington, DC: National Academy Press.

Smith, S. 1992. *Project Earth science: Astronomy.* Arlington, VA: NSTA Press.

Related Curriculum Topic Study Guides

(Keeley 2005)

"Earth, Moon, and Sun System"

"Motion of Planets, Moons, and Stars"

References

American Association for the Advancement of Science (AAAS). 1993. *Benchmarks for science literacy.* New York: Oxford University Press.

Baxter, J. 1989. Children's understanding of familiar astronomical events. *International Journal of Science Education* 11 (special issue): 502–513.

Driver, R., A. Squires, P. Rushworth, and V. Wood Robinson. 1994. *Making sense of secondary science: Research into children's ideas.* London: RoutledgeFalmer.

Keeley, P. 2005. *Science curriculum topic study:*

Bridging the gap between standards and practice. Thousand Oaks, CA: Corwin Press.

National Research Council (NRC). 1996. *National science education standards.* Washington, DC: National Academy Press.

Sadler, P. 1987. Misconceptions in astronomy. In *Proceedings of the second international seminar:* *Misconceptions and educational strategies in science and mathematics,* Vol. III, ed. J. Novak, 422–425. Ithaca, NY: Cornell University.

Schoon, K. 1992. Students' alternative conceptions of Earth and space. *Journal of Geological Education* 40: 209–214.

Emmy's Moon and Stars

Emmy looked out her window and saw the Moon and stars. She wondered how far away they were. Circle the answer that best describes where you think the Moon and stars are that Emmy sees.

A There are no stars between the Earth and the Moon.

B One star is between the Earth and the Moon.

C A few stars are between the Earth and the Moon.

D There are many stars between the Earth and the Moon.

E Several stars are between the Moon and the edge of our solar system.

Explain your thinking.

_____

_____

_____

_____

_____

_____

_____

_____

Emmy's Moon and Stars

Teacher Notes

Purpose

The purpose of this assessment probe is to elicit students' ideas about the relative position of common objects seen in the sky. The probe is designed to find out if students recognize how far away the stars are in relation to the Earth and the Moon.

Related Concepts

Earth's axis, scale size and distance in the universe, stars

Explanation

The best response is A: There are no stars between the Earth and the Moon. Even the Sun, which is the only star in our solar system, is lo-

cated far beyond the Earth and the Moon, not between it. The stars Emmy sees are located far away, outside of our solar system. To put it all in perspective, the Sun is about 150 million kilometers (93 million miles) from Earth. The next-nearest star is about 40 trillion kilometers (25 trillion miles) away. The Moon is only about 383,000 kilometers (238,000 miles) from Earth.

Distant stars, which are massive, appear as tiny points of light in the night sky because they are so far away. To a viewer on the Earth, stars may seem closer because vast distances and enormous sizes in space are difficult to visualize. Agan (2004) described the difficulty in describing stellar distance: "The vast distances

between stars are difficult for astronomers to discuss in common language. Many astronomy educators use scale models to provide a sense of the distances between stars. For instance, if the Sun were 1 inch in diameter, the nearest star would be nearly 500 miles away. A formal measurement of astronomical distances, the light year, is the distance that light travels in one year, approximately six trillion miles. The nearest star to the Sun, Proxima Centauri, is roughly 4.2 light years away."

Curricular and Instructional Considerations

Elementary Students

In the elementary years, students make regular observations of the night sky, taking inventory of the objects they see at night, including the Moon and stars. They are encouraged to draw what they see. The emphasis at this level should be on observing and describing. The magnitude of distance between these objects is beyond young children's comprehension. In addition, in the early elementary grades children lack enough of an understanding of light to realize that the brightness of the light from objects very far away, such as stars, varies according to how far away the star is. Observations and descriptions of the night sky should begin in early elementary grades with the Earth-Moon-Sun system.

In later elementary grades, students expand their observations and descriptions to

include stars and planets. Although an understanding of the location of stars in our galaxy in relation to our solar system is not a grade-level expectation, the probe is useful in identifying students' early preconceived notions, particularly because many fictional books children read at this age contribute to the development of an incorrect model of the night sky. Some of these books include illustrations that show stars in front of the Moon or nearby. The probe can be used to elicit students' ideas about where stars are located after students have developed Earth-Moon-Sun location ideas and are getting ready to learn about other objects in our solar system as well as beyond it.

Middle School Students

Students at this level begin to add details to their growing conception of the universe. The notion of scale is further developed, including much larger magnitudes and various methods and units of measurement for distant objects within and beyond our solar system. Students at this level develop a working knowledge of the apparent positions and movement of objects in the sky, including the solar system, our galaxy, and distant galaxies. Their mental models are used to explain what they see from the Earth and the positional relationships of the Earth, Moon, Sun, solar system, and beyond. This probe is useful in determining if students' earlier conceptions about where stars are located have changed as students' instructional opportunities extend beyond the solar system.

High School Students

High school is the time when a more complete picture of the vast universe develops. The study of the universe becomes more abstract. Huge magnitudes of scale make more sense to many students, although some are still at a level where abstractions and huge numbers make little sense to them. In high school, many students are ready to mathematically deal with determining greater distances. Their knowledge of physics combines with astronomy to understand how the speed of light is used to determine relative distances. This probe may be useful in determining whether high school students still retain early ideas about the location of stars, particularly if their opportunities to learn astronomical ideas have been limited.

Administering the Probe

Ask students if they have ever looked up at the sky at night and seen the Moon and the stars. Be aware that some students who live in cities may have never seen the stars. It may help to have a photograph or picture that shows the Moon and stars as they would be seen if one looked at an evening sky in a dark location. If younger students are not yet familiar with the concept of a solar system, remove distracter E or describe the solar system as the place where other planets in addition to our Moon and Sun are found.

Related Ideas in *National Science Education Standards* (NRC 1996)

. . . .

K–4 Objects in the Sky

★ The Sun, Moon, stars, clouds, birds, and airplanes all have properties, locations, and movements that can be observed.

5–8 Earth in the Solar System

• The Earth is the third planet from the Sun in a system that includes the Moon, the Sun, eight other planets and their moons, and smaller objects such as asteroids and comets. The Sun, an average star, is the central and largest body in the solar system. *[Note: This standard was written before scientists decided Pluto was not a planet.]*

Related Ideas in *Benchmarks for Science Literacy* (AAAS 1993)

. . . .

K–2 The Universe

• There are more stars in the sky than anyone can easily count, but they are not scattered evenly, and they are not all the same brightness or color.

• The Sun can be seen only in the daytime, but the Moon can be seen sometimes at night and sometimes during the day. The Sun, Moon, and stars all appear to move slowly across the sky.

3–5 The Earth

★ Stars are like the Sun, some being smaller and some larger, but so far away that they look like points of light.

• The patterns of stars in the sky stay the same, although they appear to move across

---

★ Indicates a strong match between the ideas elicited by the probe and a national standard's learning goal.

the sky nightly, and different stars can be seen in different seasons.

- Telescopes magnify the appearance of some distant objects in the sky, including the Moon and the planets. The number of stars that can be seen through telescopes is dramatically greater than can be seen by the unaided eye.

6–8 The Earth

- The Sun is a medium-sized star located near the edge of a disk-shaped galaxy of stars, part of which can be seen as a glowing band of light that spans the sky on a very clear night. The universe contains many billions of galaxies, and each galaxy contains many billions of stars. To the naked eye, even the closest of these galaxies is no more than a dim, fuzzy spot.

- ★ The Sun is many thousands of times closer to the Earth than any other star. Light from the Sun takes a few minutes to reach the Earth, but light from the next nearest star takes a few years to arrive. A trip to that star would take the fastest rocket thousands of years. Some distant galaxies are so far away that their light takes several billion years to reach the Earth. People on Earth, therefore, see them as they were that long ago in the past.

Related Research

- Students' grasp of many of the ideas of the composition and magnitude of the universe has to grow slowly over time. In spite of common depiction, the Sun-centered system seriously conflicts with common intuition (AAAS 1993).

- Agan (2004) interviewed high school and undergraduate college students to find out their ideas about distances between stars. Four out of eight high school students interviewed who had little astronomy instruction in their Earth science class and one undergraduate student out of five who received no formal astronomy instruction in high school or college described stars as being dispersed within the realm of the solar system.

Suggestions for Instruction and Assessment

- Encourage younger children to draw the objects and talk about their ideas.

- Some ideas about light and sight need to be developed before children can understand astronomical phenomena. Develop the idea early on that a large light source seen at a great distance looks like a small light source that is much closer. This phenomenon should be observed directly outside at night and, if possible, with photographs (AAAS 1993).

- Keep in mind that students' understanding of the magnitude of the universe needs to develop slowly over time. Before middle school, numbers like billions and trillions, even millions, do not make much sense to them because the vast scale is too abstract to comprehend. Even adults have difficulty comprehending how large a billion is.

- Begin teaching the notion of scale with fa-

miliar objects that students can see in the sky, such as the Moon and the Sun. Gradually introduce the nearby planets and then planets further away. Once students have grasped the enormity of our solar system, introduce the distance between Earth and our solar system, and nearby stars, gradually working outward to vast distances beyond our galaxy when students are ready to comprehend the magnitudes and measurement systems involved.

- Using telescopes—or even a good pair of binoculars—instead of the naked eye reveals more stars and makes the stars seen with the naked eye seem much brighter. Link the idea of stars being seen as points of light very far away with how telescopes help us see things, such as stars, better at significant distances. Yet, students need to be aware that the distance between Earth and the stars cannot be determined with the eyes alone or even by looking through a telescope.

- Finding distances with scale drawings helps students understand how the distances to the Moon and the Sun were estimated and why the stars must be very far away (AAAS 1993, p. 63).

Related NSTA Science Store Publications and NSTA Journal Articles

American Association for the Advancement of Science (AAAS). 1993. *Benchmarks for science literacy.* New York: Oxford University Press.

Brako, E., J. Foult, and W. Peltz. 2005. Pictures in the sky. *Science Scope* (Jan.): 49–51.

Driver, R., A. Squires, P. Rushworth, and V. Wood Robinson. 1994. *Making sense of secondary science: Research into children's ideas.* London. RoutledgeFalmer.

Gilbert, S., and S. Ireton. 2003. *Understanding models in Earth and space science.* Arlington, VA: NSTA Press.

Keeley, P. 2005. *Science curriculum topic study: Bridging the gap between standards and practice.* Thousand Oaks, CA: Corwin Press.

National Research Council (NRC). 1996. *National science education standards.* Washington, DC: National Academy Press.

Smith, S. 1992. *Project Earth science: Astronomy.* Arlington, VA: NSTA Press.

Related Curriculum Topic Study Guides

(Keeley 2005)
"Earth, Moon, and Sun System"
"Scale Size and Distance in the Universe"
"Stars and Galaxies"

References

Agan, L. 2004. Stellar ideas: Exploring students' understanding of stars. *Astronomy Education Review* 3 (1): 77–-97. Also available online at *http://aer.noao.edu/cgi-bin/article.pl?id=95.*

American Association for the Advancement of Science (AAAS). 1993. *Benchmarks for science literacy.* New York: Oxford University Press.

Driver, R., A. Squires, P. Rushworth, and V. Wood-Robinson. 1994. *Making sense of secondary*

science: Research into children's ideas. London: RoutledgeFalmer.

Keeley, P. 2005. *Science curriculum topic study: Bridging the gap between standards and practice.* Thousand Oaks, CA: Corwin Press.

National Research Council (NRC). 1996. *National science education standards.* Washington, DC: National Academy Press.

Objects in the Sky

Different things can be seen in the sky.

Put a **D** next to the things that are seen **only** in the daylight.

Put an **N** next to the things that can be seen **only** at night.

Put a **B** next to the things that can be seen in **both** day and night.

___ the Sun

___ the Moon

___ the next-nearest star to our Sun

___ constellations

Explain your thinking. How did you decide when you could see different things in the sky?

_____

_____

_____

_____

_____

_____

_____

_____

_____

Objects in the Sky

Teacher Notes

Purpose

The purpose of this assessment probe is to elicit students' ideas about when objects can be seen in the sky. Students' explanations reveal their thinking about the role of light and distance in seeing sky objects.

Related Concepts

Earth's axis, light reflection, light source, scale size and distance in the universe, stars

Explanation

The best response is D for the Sun, N for the next-nearest star to our Sun and constellations, and B for the Moon. Much to some people's surprise, the Moon can be quite visible in the blue sky when it is at a place in its orbit that puts it above the Earth's horizon during the daytime. The Moon's visibility during a bright day is due to its relative proximity to the Earth and its reflection of sunlight. A nearby star and constellations (groupings of stars) can only be seen at night because they are so far away. The only star visible to us in the daytime sky is the Sun. Venus has been called "the morning star" because of its visibility in the morning, but it is not a star. It is a nearby planet that reflects light from the sun.

Curricular and Instructional Considerations

Elementary Students

In the elementary years, students make regular observations of the sky, taking inventory of the familiar objects and their locations as seen during the day and night, including the Sun, Moon, and stars. They are encouraged to draw what they see. The emphasis at this level should be on observing and describing. Observations and descriptions of the day and night sky should begin in early elementary years with the Earth-Moon-Sun system.

In later elementary grades, students expand their observations and descriptions to include stars and planets. They also develop ideas about light reflection and light sources to explain why some things can be seen in the dark. The idea that the Moon can be seen during the daytime is a grade-level expectation in the national standards.

Middle School Students

Students at this level begin to add details to their growing conception of the solar system and the universe, moving beyond the sky overhead to the vastness of space. They develop a working knowledge of the size, distances, and movement of objects in the sky, including the planets and the effect of light reflection at a distance. They construct and use models to explain distances from the Earth and the Sun. Their growing knowledge about the motion of the Moon moves them beyond the observation that the Moon can be seen during the daytime to understanding why it can be seen based on its motion relative to the Earth and the Sun.

High School Students

High school is the time when a more complete picture of the vast universe develops and students have a more sophisticated understanding of the nature and behavior of light. Nevertheless this probe may be useful in determining whether high school students still retain early ideas about what can be seen in the sky, especially since students rarely take the time to observe these phenomena.

Administering the Probe

Listen carefully to students' ideas. You may want to probe further to ask students where they think the Moon and stars go during the daytime and where the Sun goes at night. As the inventory of celestial objects and phenomena increases for older students, you can add items to the list such as different planets, satellites, the space shuttle, the International Space Station, quasars, black holes, other moons, comets, aurora, asteroids, meteors, and meteorites.

Related Ideas in *National Science Education Standards* (NRC 1996)

. . . .

K–4 Objects in the Sky

★ The Sun, Moon, stars, clouds, birds, and airplanes all have properties, locations, and movements that can be observed.

5–8 Earth in the Solar System

• The Earth is the third planet from the Sun

★ Indicates a strong match between the ideas elicited by the probe and a national standard's learning goal.

in a system that includes the Moon, the Sun, eight other planets and their moons, and smaller objects such as asteroids and comets. The Sun, an average star, is the central and largest body in the solar system. *[Note: This standard was written before scientists decided Pluto was not a planet.]*

Related Ideas in *Benchmarks for Science Literacy* (AAAS 1993)

K–2 The Universe

★ The Sun can be seen only in the daytime, but the Moon can be seen sometimes at night and sometimes during the day. The Sun, Moon, and stars all appear to move slowly across the sky.

3–5 The Earth

• The patterns of stars in the sky stay the same, although they appear to move across the sky nightly, and different stars can be seen in different seasons.

• Planets change their positions against the background of stars.

6–8 The Earth

★ The Sun is a medium-sized star located near the edge of a disk-shaped galaxy of stars, part of which can be seen as a glowing band of light that spans the sky on a very clear night. The universe contains many billions of galaxies, and each galaxy contains many billions of stars. To the naked eye, even the closest

of these galaxies is no more than a dim, fuzzy spot.

Related Research

• Students in an astronomy class did not seem to have correct views about some astronomy-related ideas any more than students who did not have a class in astronomy. However, the students who took astronomy did use many more scientific terms in their explanations (Sadler 1987).

• The following ideas identified by Baxter (1989) may help explain where students think some sky objects are during the day or the night and why we cannot see them: the Sun goes behind hills, clouds cover the Sun, the Moon covers the Sun, the Sun goes behind the Earth once a day, the Earth goes around the Sun once a day, and the Earth spins on its axis once a day. It appears that at ages 15–16 many still hold covering and orbital theories of day and night (Driver et al. 1994).

Suggestions for Instruction and Assessment

• Provide opportunities for elementary students to observe and draw the sky at various times of the day during school hours, and encourage them to do this at night and early in the morning when they are home. Going outside and looking at the Moon during the daytime will help students see that the Moon can be visible during the day.

• Before students can discern planets in the night sky, it is necessary to help them dis-

---

★ Indicates a strong match between the ideas elicited by the probe and a national standard's learning goal.

tinguish between planets and stars in terms of both how they are seen in the sky and the difference between emitting light and reflecting it.

- Use concrete objects for models such as a ball and light. Let students observe and record how the ball looks in various locations around the light to learn how reflected light allows us to see the Moon and other planets.

- Take photographs of the sky during the day and at night or use available photographs on the internet to look at differences in the sky depending on time and season.

- Introduce students to the various types of technologies, including space telescopes, that enable us to see farther into our universe than we could with our naked eyes or land-based telescopes.

- Today's students are not personally connected to the sky as people in the past were. The sheer wonder of the sky has "inspired the expressive powers of poets, musicians, and artists" (AAAS 1993, p. 61). Help students realize that knowing the sky and what it holds is a tribute to human curiosity and our zest for understanding our place in the cosmos.

Related NSTA Science Store Publications and NSTA Journal Articles

American Association for the Advancement of Science (AAAS). 1993. *Benchmarks for science literacy.* New York: Oxford University Press.

Driver, R., A. Squires, P. Rushworth, and V. Wood Robinson. 1994. *Making sense of secondary science: Research into children's ideas.* London. RoutledgeFalmer.

Gilbert, S., and S. Ireton. 2003. *Understanding models in Earth and space science.* Arlington, VA: NSTA Press.

Keeley, P. 2005. *Science curriculum topic study: Bridging the gap between standards and practice.* Thousand Oaks, CA: Corwin Press.

National Research Council (NRC). 1996. *National science education standards.* Washington, DC: National Academy Press.

Phillips, W. 1991. Earth science misconceptions. *The Science Teacher* 58 (2): 21–23.

Smith, S. 1992. *Project Earth science: Astronomy.* Arlington, VA: NSTA Press.

Related Curriculum Topic Study Guides

(Keeley 2005)

"Earth, Moon, and Sun System"

"Scale Size and Distance in the Universe"

"Solar System"

"The Universe"

References

American Association for the Advancement of Science (AAAS). 1993. *Benchmarks for science literacy.* New York: Oxford University Press.

Baxter, J. 1989. Children's understanding of familiar astronomical events. *International Journal of Science Education* 11 (special issue): 502–513.

Driver, R., A. Squires, P. Rushworth, and V. Wood Robinson. 1994. *Making sense of secondary*

science: Research into children's ideas. London: RoutledgeFalmer.

Keeley, P. 2005. *Science curriculum topic study: Bridging the gap between standards and practice.* Thousand Oaks, CA: Corwin Press.

National Research Council (NRC). 1996. *National science education standards.* Washington, DC: National Academy Press.

Phillips, W. 1991. Earth science misconceptions. *The Science Teacher* 58 (2): 21–23.

Sadler, P. 1987. Misconceptions in astronomy. In *Proceedings of the second international seminar: Misconceptions and educational strategies in science and mathematics,* Vol. III, ed. J. Novak, 422–425. Ithaca, NY: Cornell University.

Index

National Science Teachers Association

reflection, encouraging of continuous, 9
revolution, and planetary rotation, 175
Robertson, Bill, 13
rock cycle, 153, 169
rock dust, 155
rocks
    concept matrix for probes, 150
    "Is It a Rock?" probes, 151–56, 157–62
rotation, planetary, 150, 171–75
Roth, K., 105, 110
Russell, T., 105
Ryman, D., 97–98

Sadler, Philip, x, 174
sand, 152, 154–55
scale models, 179, 182
Schneps, Matthew, x
Science Curriculum Topic Study: Bridging the Gap Between
    Standards and Practice (Keeley 2005), xi, 13
Science Formative Assessment: 75 Practical Strategies for
    Linking Assessment, Instruction, and Learning (Keeley,
    Forthcoming), 9
Science Store (NTSA), and additional materials for probes.
    See specific probes
sedimentary rock, 167, 168
seeds, 92, 101–106
sensory reasoning, 23
Shapiro, Bonnie, ix
single-celled organisms, 141
sinking and floating
    "Comparing Cubes" probe, 19–24
    concept matrix for probes, 18
    "Floating High and Low" probe, 33–39
    "Floating Logs" probe, 27–32
    "Solids and Holes" probe, 41–46
size
    of atoms, 23
    of cells, 137–42
    of universe, 150, 177–82, 185–89
Smith, C., 23
Smith, E., 105, 110
Smith, R., 74
solar system
    "Emmy's Moon and Stars" probe, 177–82
    "Objects in the Sky" probe, 185–89
"Solids and Holes" probe, 18, 22, 29–30, 36, 41–46
space science. *See* astronomy
stars
    concept matrix for probes, 150
    "Emmy's Moon and Stars" probe, 177–82
    "Objects in the Sky" probe, 185–89
station approach, and "Floating High and Low" probe, 38
Stavy, R., 140–41
Stead, B., 98
Stop Faking It! Finally Understanding Science So You Can
    Teach It series (Robertson), 10, 13
students, ideas of and use of probes, 5–7. *See also* elementary
    students; high school students; learning; middle school
    students; teaching
subtraction strategy, and temperature, 87

surface area-to-volume ratio, 139, 141

Tamir, P., 117
taxonomy. *See* biological classification
teacher notes, on use of probes, 9–13
teaching. *See also* card sort; inquiry-based investigation;
    observational experiences; station approach; students
    embedding of probes in instruction, 7–9
    linking of probes with learning and, 3–4
    vignette on topic of density, 13–15
telescopes, 182, 189
temperature
    "Boiling Time and Temperature" probe, 53–58
    concept matrix for probes, 18
    "Freezing Ice" probe, 59–64
    "Mixing Water" probe, 83–89
    "Turning the Dial" probe, 47–52
terminology. *See also* biological classification
    astronomy, 173, 175
    Earth science, 152
    life science, 116, 118, 124, 132, 134
physical science, 29, 35, 38, 70
thermal energy, 78, 85
"think-pair-share" strategy, 8
Tiberghien, Andrée, ix
time, geologic, 162, 166, 167
time-temperature graph, 57, 58
Tirosh, D., 140–41
transformation of matter, 92, 121–27
Treagust, D., 133
"Turning the Dial" probe, 18, 22, 47–52, 56, 61, 68

United Kingdom, and research on formative assessment, x
universe
    concept matrix for probes, 150
    size of, 150, 177–82, 185–89
uplift, geologic, 150, 166–70

vocabulary. *See* terminology
volume
    "Comparing Cubes" probe, 20, 24
    "Floating High and Low" probe, 35, 38
    "Floating Logs" probe, 29
    ratio of surface area to, 139, 141
    "Solids and Holes" probe, 43, 45

Wandersee, J., 110, 117, 126
water vapor, 70
Watt, D., 105
weathering and erosion, 150, 165–70
websites, 12, 13. *See also* SciLinks
weight
    "Comparing Cubes" probe, 19–24
    concept matrix for probes, 18concepts of density and
    mass, 37
    "Giant Sequoia Tree" probe, 123
"Whale and Shrew" probe, 92, 137–42
"What's in the Bubbles?" probe, 2, 18, 56, 65–70
Wiser, M., 23
Wood-Robinson, C., 133